The Best American Science Writing 2004

THE BEST AMERICAN SCIENCE WRITING

EDITORS

2000: *James Gleick*
2001: *Timothy Ferris*
2002: *Matt Ridley*
2003: *Oliver Sacks*

FORTHCOMING

2005: *Alan Lightman*

The Best American
SCIENCE WRITING
2004

EDITOR: DAVA SOBEL

Series Editor: Jesse Cohen

An Imprint of HarperCollins*Publishers*

THE BEST AMERICAN SCIENCE WRITING 2004

Permissions appear following page 267.

Compilation copyright © 2004 by HarperCollins Publishers.

Introduction copyright © 2004 by Dava Sobel.

HarperCollins books may be purchased for educational, business, or sales promotional use. For information, please write: Special Markets Department, HarperCollins Publishers Inc., 10 East 53rd Street, New York, NY 10022.

FIRST EDITION

Designed by Cassandra J. Pappas

Library of Congress Cataloging-in-Publication Data has been applied for.

ISBN 0-06-072639-3 HARDCOVER
ISBN 0-06-072640-7 TRADE PAPERBACK

04   05   06   07   08   BVG/RRD   10   9   8   7   6   5   4   3   2   1

Contents

Introduction by Dava Sobel

OVER THE PAST THIRTY-FIVE YEARS, I have worked as a newspaper reporter, a magazine writer and occasional editor, as well as a coauthor and author of books, but I always state my occupation as "science writer." I love the distinction, the nerdiness of the term and of course the science, which I loosely define as everything arcane and wonderful worth knowing, though often difficult to fathom, unpopular, and best described in mathematical terms.

I am also a science reader. I seek out excellent nonfiction that explains to me the workings of the world and its creatures, or exposes the clever inventiveness of other people, or solidifies my connection to the universe.

This volume, which I am honored to edit, is my crowning personal creation. Although I didn't write any part of it beyond this introduction, I am here empowered to affirm *The Best American Science Writing 2004*. (In truth, all the pieces appeared in 2003, the "2004" being an artifact of book publication lag.) I am happy to tie my name to the company of contributors here assembled, not to mention the four distinguished science writers who preceded me as guest editors in this yearly enterprise.

The title boasts the "best," and we—here I invoke the editorial "we" and also acknowledge the tremendous help I've received from series editor Jesse Cohen—*we* read a wealth of material and selected from it the best or, rather, what we deemed the best. On this point, I note with interest that although a rival publisher also makes available a separate and unrelated annual anthology of science writers' best work, the years pass with virtually no overlap between the two collections. What constitutes "the best science writing" is largely a matter of opinion, and only the reigning editor's opinion counts. Although the system is imperfect, subject to accidental oversight and personal preference or prejudice, I submit that every anthology as well as every literary prize, including the Pulitzer and the Nobel, is similarly filtered through a few privileged judges.

Therefore I hasten to explain my criteria for choosing only these twenty-three pieces, and rejecting the many contenders that might strike someone else as worthy of inclusion.

First and most important, all are extremely well written. This sounds obvious, and it is, but for me it means the pieces impart genuine pleasure via the writers' choice of words and the rhythm of their phrases, some of which demand immediate and repeated rereading, perhaps even reading aloud, like this bit of "A Comet's Tale" by Tom Bissell: "Apocalypticism, it hardly need be said, drags out of humanity all that is small and terrible and mean. It also drags out what is worst about God, for whom love seems an infrequent mood."

"I wish I'd written that," was my own frequent reaction to the articles I ultimately chose. Quite a few of them could just as easily be reprinted in an anthology of "best writing" and never mind the science part. But their treatment of science, whether they can be considered "science writing," is my second criterion. Of course Jesse Cohen and I searched the real science magazines from *Science* to *Discover*, with a few happy successes, but more often we found what we were looking for in the pages of *The New Yorker, The Atlantic Monthly,* and *Harper's*—where science stories appeared because of their intrinsic interest and general appeal, such as "Fun with Physics," by K. C. Cole:

> Janet Conrad fell in love with the universe at 3 AM on a cold autumn night in Wooster, Ohio. A teenager, she had no desire to get out of bed and face the frigid air in order to help her father, a dairy scientist, spray warm water on the prize dahlias they were growing together. But when she did go out to the garden she saw, for the first time in her life, how a shower of electrically charged particles flung from a star ninety-three million miles away can cover the sky in glowing pastel curtains. "I remember standing there and looking at the northern lights, and it was so neat that something so remote, so very far away, could be creating something so beautiful right in front of my eyes," she says. Twenty-five years later, Conrad, thirty-nine and an associate professor of physics at Columbia University, created her own universe—a spherical parti-

cle detector, forty feet in diameter, that she built under an igloo of dirt at the Fermi National Accelerator Laboratory (Fermilab), near Chicago. The particle detector is lined with a constellation of twelve hundred eight-inch-wide "eyes," or phototubes, and is filled with eight hundred tons of baby oil, which is used to detect the shock waves generated by particle interactions. In the fall of 2002, the detector began an unblinking vigil for subatomic stealth particles known as "sterile" neutrinos.

After "well written" and "about science," my judging criteria dissolve into fine points and specifics impossible to rank in order of importance, and some of them as vague, I admit, as the writer's attitude or tone of voice.

As much as I admired the blending of personal life with professional activity in the profile of astrophysicist Janet Conrad, I turned down numerous articles that tried but failed to achieve a similar balance. I shudder to see the wrong emphasis placed on a scientist's odd habits or on ordinary habits invoked and stressed as if to say, "Even though these people are chemists, they engage in lots of normal everyday activities!" Understatement, as per Ian Parker in "Reading Minds," is often the best policy: "Musiris was continuing to train with Palomino, who was e-mailing data to Germany each day, although the experiment had been interrupted for several weeks when Lorenzo, Estrella's pet monkey, damaged the amplifier. Lorenzo was now living with neighbors."

I am particularly hypersensitive to the way case studies are treated in medical stories. My favorite doctors in print, here represented by Sherwin Nuland and Atul Gawande, never distance themselves from the hypothetical "Jane X. (not her real name)." They have lively, respectful means of evoking those in their care, and they hasten to turn their investigations inward when necessary, as Dr. Gawande demonstrates in "Desperate Measures":

What kind of person can do this again and again—inflict suffering because of an unproved idea, a mere scientific hope? Inflicting suffering is part of any doctor's life, especially a surgeon's. We open, amputate, slice, and burn. I have put needles into screaming children who required two nurses to hold them down. I have put three-foot rubber tubes through people's noses down to their stomachs, though it made them gag and vomit and curse me in several languages. In many cases, I've found the prospect so loathsome that I looked for every excuse to avoid it. But you push ahead, because what you are doing will help the patient. You have experience and textbooks full of evidence that assure you it does. And, because it does help the patients, they forgive you; once in a while, they even thank you.

Doctors, scientists, and science writers who report for newspapers, as opposed to magazines, can hardly ever stretch the confines of their column-inch allotments for introspection or philosophical consideration. I suspect this is why few newspaper items ever find themselves reprinted as exemplars of the best science writing. But I thrill to see a working journalist surmount the challenges of harsh deadlines, short space, and the need to name every quoted source with his or her full title and affiliation (sometimes including even a previous affiliation!) while deftly exploring the intricacies of game theory, cosmology, or archaeology, à la Keay Davidson, Dennis Overbye, Tom Siegfried, and John Noble Wilford.

As I read through the year in science, I thought ahead to the eventual shape this collection would take and what grand purpose it might serve. Individually, every item was once discarded with yesterday's news, but, gathered here between hard covers, the pieces can't help but portray the recent history and development of science.

The year 2003 opened on the fatal reentry of the space shuttle *Columbia*. The February 1 tragedy, along with the investigation into the nationwide complacency that brought the ship down, is brilliantly recounted by William Langewiesche in "*Columbia*'s Last Flight":

> The astronauts aboard the shuttle were smart and accomplished people, and they were deeply committed to human space flight and exploration. They were also team players, by intense selection, and nothing if not wise to the game. From orbit one of them had radioed, "The science we're doing here is great, and it's fantastic. It's leading-edge." Others had dutifully reported that the planet seems beautiful, fragile, and borderless when seen from such altitudes, and they had expressed their hopes in English and Hebrew for world peace. It was Miracle Whip on Wonder Bread, standard NASA fare. On the ground so little attention was being paid that even the radars that could have been directed upward to track the *Columbia*'s re-entry into the atmosphere—from Vandenberg Air Force Base, or White Sands Missile Range—were sleeping. As a result, no radar record of the breakup exists—only of the metal rain that drifted down over East Texas, and eventually came into the view of air-traffic control.

After selecting all the articles, I considered arranging them alphabetically by author, except that Jesse Cohen directed me to discern some other order among them. What could have been a simple matter of filing thus turned into an organizational puzzle. I chose Jennifer Kahn's "Stripped for Parts" as the lead story because it begins with one of the most arresting openings I have ever read:

> The television in the dead man's room stays on all night. Right now the program is *Shipmates,* a reality-dating drama that's barely audible over the hiss of the ventilator. It's 4 AM, and I've been here for six hours, sitting in the corner while three nurses fuss intermittently over a set of intravenous drips. They're worried about the dead man's health.

Closing the collection with Diane Ackerman's ebullient nature essay was also an obvious, easy choice, so as to end on a joyful note. In between these anchors, I weighed the topic and mood of each piece, the possible tension between bacterium and dark energy, the varying lengths of articles, the fact that one item for sure and arguably two are poems, and the contrast of timeless essay with breaking news.

Having imposed the present order for my own reasons, I encourage you to ignore it. Feel free to start with your own favorite writer or subject and proceed as randomly as you wish—the perfect way, I think, to navigate an anthology in which every entry always meant to stand on its own.

The Best American Science Writing 2004

JENNIFER KAHN

Stripped for Parts

FROM *WIRED*

The television in the dead man's room stays on all night. Right now the program is *Shipmates*, a reality-dating drama that's barely audible over the hiss of the ventilator. It's 4 AM, and I've been here for six hours, sitting in the corner while three nurses fuss intermittently over a set of intravenous drips. They're worried about the dead man's health.

To me, he looks fine. His face is slack but flush, he breathes steadily, and his heart beats like a clock, despite the fact that his lungs have recently begun to leak fluid. The nurses roll the body from side to side periodically so that the liquid doesn't pool. At one point, a white plastic vest designed to clear the lungs inflates and begins to vibrate violently—as if some invisible person has seized the dead man by the shoulders and is trying to shake him awake. The rest of the time, the nurses consult monitors and watch for signs of cardiac arrest. When someone scratches the bottom of the dead man's foot, it twitches.

None of this is what I expected from an organ transplant. When I arrived last night at this Northern California hospital I was prepared to see a fast-paced surgery culminating in renewal: the mortally ill patient restored to glorious health. In all my preliminary research on transplants, the dead man was rarely mentioned. Even doctors I spoke with avoided the subject, and popular accounts I came across ducked the matter of provenance altogether. In the movies, for instance, surgeons tended to say it would take time to "find" a heart—as though one had been hidden behind a tree or misplaced along with the car keys. Insofar as corpses came up, it was only in anxious reference to the would-be recipient whose time was running out.

In the dead man's room, a different calculus is unfolding. Here the organ is the patient, and the patient a mere container, the safest place to store body parts until surgeons are ready to use them. It can be more than a day from the time a donor dies until his organs are harvested—the surgery alone takes hours, not to mention the time needed to do blood tests, match tissue, and fly in special surgical teams for the evisceration. And yet, a heart lasts at most six hours outside the body, even after it has been kneaded, flushed with preservatives, and packed in a cooler. Organs left on ice too long tend to perform poorly in their new environment, and doctors are picky about which viscera they're willing to work with. Even an ailing cadaver is a better container than a cooler.

These conditions create a strange medical specialty. Rather than extracting this man's vitals right away, the hospital contacts the California Transplant Donor Network, which dispatches a procurement team to begin "donor maintenance": the process of artificially supporting a dead body until recipients are ready. When the parathyroid gland stops regulating calcium, key to keeping the heart pumping, the team sends the proper amount down an intravenous drip. When blood pressure drops, they add vasoconstrictors, which contract the blood vessels. Normally the brain would compensate for a decrease in blood pressure, but with it out of commission, the three-nurse procurement team must take over.

In this case, the eroding balance will have to be sustained for almost 24 hours. The goal is to fool the body into believing that it's alive and well, even as everything is falling apart. As one crew member concedes, "It's unbelievable that all this stuff is being done to a dead person."

Unbelievable and, to me, somehow barbaric. Sustaining a dead body until its organs can be harvested is a tricky process requiring the latest in medical technology. But it's also a distinct anachronism in an era when medicine is becoming less and less invasive. Fixing blocked coronary arteries, which not long ago required prying a patient's chest open with a saw and spreader, can now be accomplished with a tiny stent delivered to the heart on a slender wire threaded up the leg. Exploratory surgery has given way to robot cameras and high-resolution imaging. Already, we are eyeing the tantalizing summit of gene therapy, where diseases are cured even before they do damage. Compared with such microscale cures, transplants—which consist of salvaging entire organs from a heart-beating cadaver and sewing them into a different body—seem crudely mechanical, even medieval.

"To let an organ reach a state where the only solution is to cut it out is not progress; it's a failure of medicine," says pathologist Neil Theise of NYU. Theise, who was the first researcher to demonstrate that stem cells can become liver cells in humans, argues that the future of transplantation lies in regeneration. Within five years, he estimates, we'll be able to instruct the body to send stem cells to the liver from the store that exists in bone marrow, hopefully countering the effects of a disease like hepatitis A or B and letting the body heal itself. And numerous researchers are forging similar paths. One outspoken surgeon, Richard Satava from the University of Washington, says that medicine is only now catching on to the fundamental lesson of modern industry, which is that when our car alternator breaks, we get a brand new one. Transplantation, he argues, is a dying art.

Few researchers predict that human-harvested organs will become obsolete anytime soon, however; one cardiovascular pathologist, Charles Murry, says we'll still be using them a century from now. But it's reasonable to expect—and hope for—an alternative. "I don't think anybody enjoys recovering organs," Murry says frankly. "You tell yourself it's for a good cause, which it is, a very good cause, but you're still butchering a human."

INTENSIVE CARE IS NOT a good place to spend the evening. Tonight, the ward has perhaps 12 patients, including a woman who moans constantly and a deathly pale man who reportedly jumped out the window of a moving Greyhound bus. The absence of clocks and the always-on lights create a casino-like timelessness. In the staff lounge, which smells of stale pizza, a lone nurse corners me and describes watching a man bleed to death ("He was conscious. He knew what was happening"), and announces, sotto voce, that she knows of South American organ brokers who charge $60,000 for a heart, then swap it for a baboon's.

Although I don't admit it to the procurement team, I've grown attached to the dead man. There's something vulnerable about his rumpled hair and middle-aged body, naked save a waist-high sheet. Under the hospital lights, everything is exposed: the muscular arms gone flabby above the elbow; the legs, wiry and lean, foreshortened under a powerful torso. It's the body of a man in his fifties, simultaneously bullish and elfin. One foot, the right, peeps out from the sheet, and for a brief moment I want to hold it and rub the toes that must be cold—a hopeless gesture of consolation.

Organ support is about staving off entropy. In the moments after death, a cascade of changes sweeps over the body. Potassium diminishes and salt accumulates, drawing fluid into cells. Sugar builds up in the blood. With the pituitary system offline, the heart fills with lactic acid like the muscles of an exhausted runner. Free radicals circulate unchecked and disrupt other cells, in effect causing the body to rust. The process quickly becomes irreversible. As cell membranes grow porous, a "death gene" is activated and damaged cells begin to self-destruct. All this happens in minutes.

When transplant activists talk about an organ shortage, it's usually to lament how few people are willing to donate. This is a valid worry, but it eclipses an important point, which is that the window for retrieving a viable organ is staggeringly small. Because of how fast the body degrades once the heart stops, there's no way to recover an organ from someone who dies at home, in a car, in an ambulance, or even while on the operating table. In fact, the only situation that really lends itself to harvest is brain death, which means finding an otherwise healthy patient whose brain activity has ceased but whose heart continues to beat—right up until the moment it's taken out. In short, victims of stroke or severe head injury. These cases are so rare (approximately 0.5 percent of all deaths in the US) that even if everybody in America were to become a donor, they wouldn't clear the organ wait lists.

This is partly a scientific problem. Cell death remains poorly understood, and for years now, cadaveric transplants have lingered on a research plateau. While immunosuppressants have improved incrementally, transplants proceed much as they did 20 years ago. Compared with a field like psychopharmacology, the procedure has come to a near-standstill.

But there are cultural factors as well. Medicine has always reserved its glory for the living. Even among transplant surgeons, a hierarchy exists: Those who put organs into living patients have a higher status than those who extract them from the dead. One anesthesiologist confesses that his peers don't like to work on cadaveric organ recoveries. (Even brain-dead bodies require sedation, since spinal reflexes can make a corpse "buck" in surgery.) "You spend all this time monitoring the heartbeat, the blood pressure," the anesthesiologist explains. "To just turn everything off when you're done and walk out. It's bizarre."

Although the procurement team will stay up all night, I break at 4:30 AM for a two-hour nap on an empty bed in the ICU. The nurse removes a wrinkled top sheet but leaves the bottom one. Doctors sleep like this all the time, I know, catnapping on gurneys, but I can't shake the feeling of climbing onto my deathbed. The room is identical to the one I've been sitting in for the past eight hours, and I'd prefer to sleep almost anywhere else—in the nurses lounge or even on the small outside balcony. Instead, I lie down in my clothes and pull the sheet up under my arms.

For a while I read a magazine, then finally close my eyes, hoping I won't dream.

By morning, little seems to have changed, except that the commotion of chest X-rays and ultrasounds has left the dead man's hair more mussed. On both sides of his bed, vital stats scroll across screens: oxygen ratios, pulse, blood volumes.

All of this vigilance is good, of course: After all, transplants save lives. Every year, thousands of people who would otherwise die survive with organs from brain-dead donors; sometimes, doctors say, a patient's color will visibly change on the operating table once a newly attached liver begins to work. Still—and with the possible exception of kidneys—transplants have never quite lived up to their initial promise. In the early 1970s, few who received new organs lasted even a year, and most died within weeks. Even today, 22 percent of heart recipients die in less than four years, and 12 percent reject a new heart within the first few months. Those who survive are usually consigned to a lifetime regime of costly immunosuppressive drugs, some with debilitating side effects. Recipients of artificial hearts traditionally fare the worst, alongside those who receive transplants from animals. Under the circumstances, it took a weird kind of perseverance for doctors operating in 1984 to suggest sewing a walnut-sized baboon heart into a human baby. And there was grief, if not surprise, when the patient died of a morbid immune reaction just 21 days later.

BY THE TIME we head into surgery, the patient has been dead for more than 24 hours, but he still looks pink and healthy. In the operating room, all the intravenous drips are still flowing, convincing the body that everything's fine even as it's cleaved in half.

Although multiorgan transfer can involve as many as five teams in the OR at once, this time there is only one: a four-man surgical unit from Southern California. They've flown in to retrieve the liver, but because teams sometimes swap favors, they'll also remove the kidneys for a group of doctors elsewhere—saving them a last-minute, late-night flight. One of the doctors has brought a footstool for me to stand on at the head of the operating table, so that I can see over the sheet that hangs between the patient's head and body. I've been warned that the room will smell bad during the "opening," like flesh and burning bone—an odor that has something in common with a dentist's drill.

Behind me, the anesthesiologist checks the dead man's mask and confirms that he's sedated. The surgery will take four hours, and the doctors have arranged for the score of Game Five of the World Series to be phoned in at intervals.

I've heard that transplant doctors are the endurance athletes of medicine, and the longer I stand on the stool, the better I understand the comparison. Below me, the rib cage has been split, and I can see the heart, strangely yellow, beating inside a cave of red muscle. It doesn't beat forward, as I expect, but knocks anxiously back and forth like a small animal trapped in a cage. Farther down, the doctors rummage under the slough of intestines as though through a poorly organized toolbox. When I tell the anesthesiologist that the heart is beautiful, he says that livers are the transplants to watch. "Hearts are slash and burn," he shrugs, adjusting a dial. "No finesse."

Two hours pass, and the surgeons make progress. Despite the procurement team's best efforts, however, most of the organs have already been lost. The pancreas was deemed too old before surgery. One lung was bad at the outset, and the other turned out to be too big for the only matching recipients—a short list given the donor's rare blood type. At 7 this morning, the heart went bust after someone at the receiving hospital suggested a shot of thyroid hormone, shown in some studies to stimulate contractions—but even before then, the surgeon had had second thoughts. A 54-year-old heart can't travel far—and this one was already questionable—but the hospital may have thought this would improve its chances. Instead, the dead man's pulse shot to 140, and his blood began circulating so fast it nearly ruptured his arteries. Now the heart will go to Cryolife, a biosupply company that irradiates and freeze-dries the valves, then packages them for sale to hospitals in screw-top jars. The kidneys have remained healthy enough to be passed on—one to a man who will soon be in line for a pancreas, the other to a 42-year-old woman.

Both kidneys have been packed off in quart-sized plastic jars. Originally, the liver was going to a nearby hospital, but an ultrasound suggested it was hyperechoic, or probably fatty. On the second pass, it was accepted by a doctor in Southern California and ensconced in a bag of icy slurry.

The liver is enormous—it looks like a polished stone, flat and purplish—and with it gone, the body seems eerily empty, although the heart continues to beat. Watching this pumping vessel makes me oddly anxious. It's sped up slightly, as though sensing what will happen next. Below me, the man's face is still flushed. He's the one I wish would survive, I realize, even though there was never any chance of that. Meanwhile, the head surgeon has walked away. He's busy examining the liver and relaying a description over the phone to the doctor who will perform the attachment. Almost unnoticed, an aide clamps the arteries above and below the heart, and cuts. The patient's face doesn't move, but its pinkness drains to a waxy yellow. After 24 hours, the dead man finally looks dead.

Once all the organs are out, the tempo picks up in the operating room. The heart is packed in a cardboard box also loaded with the kidneys, which are traveling by Learjet to a city a few hundred miles away. Someday, I'm convinced, transporting organs in coolers will seem as strange and outdated as putting a patient in an iron lung. In the meantime, transplants will survive: a vehicle, like the dead man, to get us to a better place. As an assistant closes, sewing up the body so that it will be ready for its funeral, I get on the plane with the heart and the kidneys. They've become a strange, unhealthy orange in their little jars. But no one else seems worried. "A kidney almost always perks up," someone tells me, "once we get it in a happier environment."

ATUL GAWANDE

Desperate Measures

FROM THE NEW YORKER

On November 28, 1942, an errant match set alight the paper fronds of a fake electric-lit palm tree in a corner of the Cocoanut Grove night club near Boston's theatre district and started one of the worst fires in American history. The flames caught onto the fabric decorating the ceiling, and then swept everywhere, engulfing the place within minutes. The club was jammed with almost a thousand revellers that night. Its few exit doors were either locked or blocked, and hundreds of people were trapped inside. Rescue workers had to break through walls to get to them. Those with any signs of life were sent primarily to two hospitals—Massachusetts General Hospital and Boston City Hospital. At Boston City Hospital, doctors and nurses gave the patients the standard treatment for their burns. At MGH, however, an iconoclastic surgeon named Oliver Cope decided to try an experiment on the victims. Francis Daniels Moore, then a fourth-year surgical resident, was one of only two doctors working on the emergency ward when the victims came in. The experience, and the experiment, changed him. And, because they did, modern medicine would never be the same.

It had been a slow night, and Moore, who was twenty-nine years old, was up in his call room listening to a football game on the radio. At around 10:30 PM, he heard the familiar whine of an ambulance arriving outside, put on his white coat, and went to see what was going on. Making his way to the ward, he heard another ambulance arriving—then another, and another. He broke into a run. In less than two hours, he received a hundred and fourteen burn victims. He described the scene several days later in a letter to his parents:

> Down the hall were streaming stretchers with burned people on them. One a young girl, with her clothing burned off, and her skin hanging like ribbons as she flailed her arms around, screaming with pain. Another a naval lieutenant who kept repeating over and over again, "I must find her. I must find her." His face and hands were the dead paper-white that only a deep third-degree burn can be, and I knew only looking at him for a moment that if he lived, in two weeks his face would be a red, unrecognizable slough. He didn't live.

Moore grabbed a syringe full of morphine and gave anyone he found alive a slug for the pain. Dozens died in those first few hours. Many succumbed from shock and overwhelming injuries. Others, some without a single burn on them, died of asphyxia, their singed throats slowly swelling closed. In all, nearly five hundred people died from the fire. Of MGH's hundred and fourteen patients, only thirty-nine survived long enough to be admitted to the hospital and treated for their burns. Bodies were laid in rows along a corridor; a hospital floor was cleared for the survivors. And, at Oliver Cope's insistence, the experiment began.

The conventional treatment for severe burns was to tan the burn surface as quickly as possible, and at Boston City that is what the surgeons did. People who initially survive bad burns remain at high risk of dying from infection in the days to follow. Your skin protects you from the germs of the outside world; a burn opens the portals. Applying tannic acid to a burn was a way to create a thickened, protective cover. Patients were given morphine and soaked in a bathtub; then their blisters were cut off and the acid slowly poured on their wounds. The process was extraordinarily painful and laborious, and sometimes fatal. It also took four or five trained personnel to care for one patient. Still, it was a proved therapy, and it had been standard practice for years.

Medicine, especially surgery, is a conservative profession; a physician departs only reluctantly from the established techniques and lessons. And for good reason: the stakes, if you are wrong, are too high. Doctors are expected to adopt new treatments only with strong evidence that they will have better results. But Cope was a believer; and one of the things he believed was that tannic-acid treatment was no good.

Earlier that year, he'd been called on for advice following the Japanese bombing of Pearl Harbor. Investigators had found that the major casualties were not from blast injuries but from burns from the fires that followed. Medical personnel had been overwhelmed by the labor required by tannic-acid treatment. A day and a half after the attack, they still had not completed the initial care for victims. Many patients died waiting. Cope proposed wrapping people's burns in gauze coated with petroleum jelly, and then leaving them alone. The treatment would be far less painful, and a single doctor could care for four or five patients by himself.

To many, the notion that a thin layer of gauze smeared with Vaseline would stop infection seemed foolish. Yet on the night of the Cocoanut Grove fire—despite numerous lives at stake, intense media attention, criticism from the surgical establishment, and the fact that MGH's results could be compared very easily with Boston City's—Cope insisted on his experimental treatment. His experience with it was almost laughably slim: he had tried it just twice (once on himself). But he had found in the laboratory that blisters protected by petroleum-jelly gauze appeared to stay sterile.

Francis Moore was intelligent and ambitious; he was also, before the fire, still relatively unformed. He had received his undergraduate and medical degrees at Harvard, but distinguished himself more as a wit than as a scholar. Chronic asthma from childhood had kept him out of the war. He was no better skilled in the operating room than his fellow-residents. But the Cocoanut Grove disaster became his Omaha Beach; it exposed him to a magnitude of suffering he scarcely imagined possible, leaving him with a thickened, protective hide of his own. And it gave him his first true expertise—he saw more serious burns from this one catastrophe than most surgeons do in their careers, and he came to publish dozens of papers in the field. Despite the range of his eventual influence, he called himself a burn surgeon the rest of his life.

Moore—or Franny, as everyone called him—went on to become one of the most important surgeons of the twentieth century. He discovered the chemical composition of the human body, and was a pioneer in the development of nuclear medicine. As the youngest chairman of surgery in Harvard's history, he led his department to attempt some of the most daring medical experiments ever conducted—experiments that established, among other things, organ transplantation, heart-valve surgery, and the use of hormonal therapy against breast cancer. Along the way, the line between patients and experimental subjects was blurred; his attempts to develop new procedures inevitably cost lives as well as saved them. His advances made medicine more radical, more invasive of human bodies, and more dependent upon technology. But in November of 1942 he got his first sense of what might be possible when you put aside custom and convention.

In the hours and days after the fire, Moore carried out Cope's experimental treatment. A month later, investigators from the National Research Council arrived in Boston. At Boston City Hospital, the council's final report said, some thirty per cent of the initial survivors had died, most from infections and other complications of their burns. At MGH, none of the initial survivors had died from their burn wounds. Cautious experience lost. And, as Moore would always remember, experiment won.

MOORE'S EXPERIMENTS PUSHED medicine harder and farther than almost anyone had contemplated—and uncomfortably beyond what contemporary medical ethics would permit. Moore did not think like a surgeon; he wasn't enthralled by technique. Doing burn research with Cope, he found his strongest interest was science.

A few months after the fire, Moore took a year off from operating to do research, enrolling in nuclear-physics courses and joining a biophysics lab. In that one year in the lab, he and a young chemist, Lester Tobin, invented a method of using radioactive tracers and a Geiger counter to detect abscesses that were invisible to X-rays. The technique obviated exploratory surgery for a range of diseases, anticipating the field known as nuclear medicine. Another time, on rounds one day when he was a resident, he came up with an operation to cure ulcers. He had been seeing a patient with stress-related ulcers and proposed that severing certain nerves to the stomach could cure the problem. A year later, as a new member of the MGH surgical faculty, he performed the operation and proved it—only to find that a doctor from the University of Chicago had beaten him by just a few months. The operation became the standard treatment for peptic ulcers for the next twenty-five years.

Moore's range and creativity were astounding. "The mind could simply overpower," said Steve Rosenberg, who trained under him and is now chief of surgery at the National Cancer Institute. Moore had a gravelly, throaty voice and a booming laugh, and he was by nature an actor. As an undergraduate, he had been president of both the *Harvard Lampoon* and the Hasty Pudding Theatricals. He had written plays, composed musicals, and, with his friend Alistair Cooke, once put together a show entitled "Hades! The Ladies!," which they took on the road, performing it in the Roosevelt White House. He never lost the ability to command an audience, small or large. "He had that attractive *pause*," Leroy Vandam, an anesthesiologist who worked alongside him for decades, said.

In 1945, in an experiment that made Moore's career, he tackled a seemingly inconsequential but long-unanswered question: How much water is in the human body? Even into the nineteen-forties, nobody had a precise idea. Scientists knew that people were made of salts and minerals and water and fats and so on in some kind of balance and distribution, but they did not know much more than that or how they might find out. Moore hit upon an idea while sitting, Archimedes-like, in his bathtub: Suppose he put a drop of heavy water in the tub water, let it diffuse, and then took a drop of the tub water out. If he measured how much the deuterium had been diluted, he'd be able to calculate exactly how much water was in the tub. And he could do the same thing in a human being.

A professor he'd met while doing isotope research sent Moore a test tube of radioactive deuterium from a nuclear reactor. (These were looser times.) Working in the small laboratory he'd been given as a junior faculty member, he injected a few millilitres of the heavy water into some rabbits, let it equilibrate over an hour, then withdrew samples of blood by vein. He measured the concentration of deuterium using a densitometer and calculated each rabbit's total body-water volume. Then he measured the amount of water in the rabbits directly, by drying them in a vacuum-desiccator. His calculations, it turned out, were perfect.

Moore began doing heavy-water studies in human beings. He tested almost anyone he could get his hands on: laboratory workers, their families, and many others. He tried other radioactive isotopes, too—of sodium, potassium, iron, phosphorus, chromium—and gradually figured out the amounts and behaviors of the essential ingredients of the human body. Then he and a growing team of researchers began taking measurements of patients with traumatic wounds, chronic illnesses, massive hemorrhages. The work paid off: study by study, the doctors opened the black box of human chemistry.

Here are some of the things Moore discovered. That the average adult male is fifty-five per cent water and the adult female is fifty per cent. That this difference appears during puberty, when boys lose fat and girls gain it: muscle sequesters water and fat displaces it, so that a very skinny person can be as much as seventy per cent H_2O. That in acute illness water leaks out of blood vessels into the soft tissues, sometimes by the litre; the heart tries to adjust by increasing its rate; and the whole process is controlled by hormones in the bloodstream. That muscle protein is burned for energy, and this is why sick people waste away.

He also learned that critical illness often arose from a simple chemical imbalance—that, for instance, what often killed patients with vomiting or severe infection was a lack of potassium, and that, with carefully calibrated infusions of salt water and potassium, these deaths could be prevented. He showed that in some patients one could correct delirium, convulsions, or heart-rhythm disturbances with small amounts of magnesium, and that sick patients require particular proteins and minerals and sugars in order to survive. Through more than a decade of studies, he worked out, for disease after disease, a step-by-step process for getting people well.

The revolutionary nature of Moore's work was recognized almost immediately. In July, 1948, at the age of thirty-four, he was made full professor and chairman of surgery at Harvard Medical School's Peter Bent Brigham Hospital. In 1959, he compiled his findings in what became one of the top-selling medical textbooks ever.

In dozens of ways, Moore's research simply made medicine better—more informed, more systematic, more effective. He had made a science of convalescence. Surgery, in particular, changed dramatically. Even the most elementary operations had carried substantial risk of death—the average person stood a one-in-twenty chance of dying from an appendectomy, an operation that some seven per cent of human beings require. But Moore showed surgeons how to make a range of small adjustments—from how to titrate fluids and salts to how to compensate for bleeding or shock—that made routine surgery safer by orders of magnitude. His findings probably saved tens of thousands of lives a year—an impact on the scale of vaccines and antibiotics.

At the same time, Moore's work, by lowering the risk of invading the human body, gave doctors the confidence to try things that would once have seemed outlandish. And, in this, no one was more ambitious than Francis D. Moore.

I KNEW DR. MOORE as a formidable though distant presence in my surgical training. The hospital where he was chairman of surgery is my hospital—it is now called Brigham and Women's Hospital—and though he was long retired when my residency began, a few years ago, he still attended weekly surgical conferences. You always shook a little when you took the podium to explain a patient's case and looked out to find him sitting in the front row. White-haired to his eyebrows, slouching a bit in his seat, his hands folded casually, he hardly appeared threatening. Yet when the time came for his comments he seemed to know everything about the science and the surgery and the saving of human lives, and he didn't hesitate to make his views known.

In an operating room, he could be chatty and make little jokes. But even when he was in his thirties, with staff sometimes decades older than he, the room was firmly in his command. He liked to whistle during cases, for example. (Some surmised that he did so to control his wheezy, asthmatic breathing—he had a poorly concealed cigarette habit that had turned his childhood lung troubles into emphysema, but he could whistle an entire Brandenburg concerto during an operation without dropping a note.) And, when he did, no one made a sound.

Although he was not an especially fast or elegant surgeon, he was meticulous, methodical, and, when he needed to be, inventive. He didn't like surprises in operations, but he came prepared for them. Some surgeons arrive in the operating room with only a vague idea of what they will do. Faced with a nonroutine problem—an unusually massive hernia, say, or a colon cancer invading the spleen—they improvise, relying on instinct and experience. Moore thought hard about each case ahead of time. Once he took a patient to the operating room, he had decided on every move he wanted to make.

Like many attending surgeons, he used to browbeat the residents who operated with him. But his concern was not their technique. "He really wanted to know how you thought," Gordon Vineyard, a Boston surgeon who had trained under him, told me. "He wanted to know what you were going to do. How. Why. Especially why. He got in your head." People had put their lives in your hands, so you'd better know everything you could about them. "He'd ask you about a lab test for a patient, and, if you didn't know the result, that was a moral failure," Vineyard said. "But if you ever guessed, or made it up, you were *dead.*"

"The fundamental act of medical care is assumption of responsibility . . . complete responsibility for the welfare of the patient," Moore wrote on the first page of his textbook. A good doctor, he went on, "employs any effective means available." And if there is no effective means available? Then you must try to come up with one. Death, he argued, must never be seen as acceptable. Confronted with a dying patient, he did not hesitate to consider the most outrageous proposals.

He saw, for example, scores of patients with advanced, metastatic breast cancers. The women invariably died, most within a year and in terrible pain, especially once the tumor had spread to their bones. Moore was troubled that medicine had given up on these women; it was his ethical and scientific judgment that no problem was beyond trying to solve. He established a team of physicians dedicated to treating these patients, and began to study the increasing evidence that certain hormones played a role in promoting breast cancer. In a radical 1955 experiment, Moore and a neurosurgeon named Don Matson began operating on the metastatic-breast-cancer patients to remove their pituitary gland—a whitish, peanut-size nerve center in the brain which controls the production of everything from estrogen to human growth hormone. The operation would shut down their endocrine systems and thereby, Moore hoped, eliminate a major stimulus of cancer growth.

Moore and his colleagues had no animal studies to point to, and the operation was tremendously risky: it required entering the skull above the right eye, exposing the frontal lobe of the brain, removing a portion just to get in far enough, then working over to the part of the brain behind the top of the nose where the pituitary dangled from its stalk. Of their first fifty-three patients, one lost the right half of her vision; several lost their sense of smell; three had major strokes; seven developed seizures; and one died. All had to go on medications to maintain minimum levels of essential hormones. Many physicians thought the whole project was insane.

Slowly, however, surprising results emerged. One patient was a forty-two-year-old woman with two children and a cancer of her right breast which had not only spread to her bones but formed thick red deposits all over her skin. She was bedridden and failing rapidly. After the operation, her skin tumors began melting away. She returned to a life at home, living in remission for fourteen months. In others, the rib and spinal metastases shrank, and their pains disappeared.

Many patients saw no benefits at all. Some lives were clearly made worse. But sixty per cent of the women improved or survived longer, and a few were actually cured—a feat that no other medical treatment had ever accomplished. With further experimentation, Moore and another colleague, Richard Wilson, found that removing the ovaries and adrenal glands (where estrogen and other hormones are produced) had equal benefits with less risk. Other physicians demonstrated similar results, and a search began for a medicine that could provide the same hormonal effects. These efforts produced the anti-estrogen drug tamoxifen, which is now a central element in both the treatment and the prevention of breast cancer.

Moore did not hesitate to expose his patients to suffering and death if a new idea made scientific sense to him. When he became chairman of surgery, one of his earliest recruits was Dwight Harken, who, during the Second World War, had performed a hundred and thirty-four successful operations to remove bullets and shell fragments from inside the hearts of wounded soldiers. The heart had long been regarded as impossible surgical territory. But Harken was young and brash, and, through trial and error, he accumulated more experience at opening the heart than anyone else in the world. As soon as Moore brought him in, Harken began attempting a dangerous operation called a mitral valvulotomy. The mitral valve is one of the four major valves of the heart, and insures that blood flows forward through the pump, instead of jetting backward. In patients with mitral stenosis (which is common after rheumatic fever), the valve becomes hardened and calcified; getting blood through it can be like trying to push a river through a straw. The afflicted die by degrees, their lungs swollen with fluid, their lips turning blue, their breathing increasingly labored. In the nineteen-twenties, surgeons had attempted opening the valve in several patients by using, first, a metal punch-hole instrument, and then simply a gloved index finger to crack through. But the results were so dismal—five of seven died from complications—that the operation was abandoned.

Moore and Harken, however, took encouragement from the fact that two people had survived. Harken's initial results were hardly an improvement. Four of his first six patients died on the table. Criticism from the medical community mounted. The surgical residents at the hospital, who had to assist in these operations and care for the patients afterward, met privately with Moore to demand a moratorium on the procedure. Even Harken began to have doubts. Still, Moore was insistent that he go on, and kept the residents in line. He pushed Harken to select healthier mitral patients for surgery, arguing that the other patients might have been too ill to survive. Perhaps doing an unproved, deadly operation was permissible as a last-ditch effort for people with only days of life remaining. But Moore wanted to operate on patients who had many months ahead of them. Harken took his advice, and gradually figured out how to improve the survival rates. In 1964, he published a complete series of more than fifteen hundred of these operations. The survival rate for those with end-stage disease was eighty-three per cent. For patients with less advanced disease, it was a stunning 99.4 per cent. The era of open-heart surgery was born.

How could you be sure that a new procedure would one day prove its worth? You couldn't. During the same period, in the nineteen-fifties, Moore led a surgical team that removed people's adrenal glands in an effort to treat severe hypertension of unknown cause. Eventually, the operation had to be abandoned; the death rate was too high, and there was little evidence that the procedure helped. But Moore's confidence in his instincts was never more bitterly tested than during his department's experiments with organ transplantation.

The taking of body parts from one being for another had a long and unsavory history. In the eighteenth century, for example, a British surgeon demonstrated that human teeth could be transplanted, and this became briefly popular in England. But it proved to be a disaster. Teeth were stolen from corpses and purchased from the poor for the gums of the gap-toothed wealthy. And though the transplanted teeth sometimes took, they also brought with them infections, including syphilis, that spread unhindered through people's jaws. In France just before the First World War, two attempts at kidney transplantation failed. Then, between the wars, there was a vogue for testicular transplants, which were thought to produce sexual "rejuvenation." In Paris between 1921 and 1926, a Russian surgeon named Serge Voronoff transplanted wedges of monkey testis into close to a thousand men. In Kansas, by 1930, J. R. Brinkley reported having performed sixteen thousand glandular transplants. Laboratory studies found, however, that within days of transplantation the grafts were destroyed by the host's rejection response. Finally, the medical profession had had enough, and shut down outfits like Brinkley's. By the nineteen-fifties, when Moore got involved, the notion of taking organs from one person and putting them into another was widely regarded as disreputable.

To Moore, though, it combined the frisson of the daring and a moral attraction: he became interested in transplantation as a way to save the lives of those whose kidneys had failed. Kidney failure—from infection, shock, the slow damage of high blood pressure—had invariably fatal consequences. Unable to excrete water or the potassium and acids that build up in the body, its victims either drowned in their own fluids or were poisoned by their own blood. It was a condition more dreaded than cancer, and medicine had nothing to offer.

In 1951, Moore, then thirty-nine, and George Thorn, the medical chief of staff, put together a team under a young surgeon named David Hume to try giving these patients kidneys from dead people. Their plans were, in retrospect, woefully primitive. Despite overwhelming evidence that the immune system would reject any foreign organ, they had no strategy for stopping the rejection process. And although they had experimented with the operation in dogs, they had not yet worked out how best to hook up a new kidney. (Should you connect it to the vessels and ureter of one of the failed kidneys, or put it somewhere else? Should you keep it warm or cold until you attached it?)

Nonetheless, by 1953, ten patients had undergone the experimental operation. The results were abysmal. The first patient had been a comatose twenty-nine-year-old woman whose kidneys had shut down owing to a hemorrhage following an illegal abortion. Hume found a hospital worker who let the doctors take a kidney from a relative who had just died. The woman was unable to travel, so Hume and his team did the operation at her bedside, late at night under a gooseneck lamp, the coma providing her anesthesia. They connected the kidney to blood vessels in her arm and left the organ outside the skin under rubber sheeting. Urine began dripping out of the exposed ureter almost immediately, and the next day she awoke from her coma. Within forty-eight hours, however, rejection took over. The kidney became massively swollen, ceased producing urine, and had to be disconnected. Surprisingly, the procedure gave her just enough time for her own kidneys to recover, but she died not long after from hepatitis contracted from the blood transfusions she had required. The subsequent patients did little better. Every transplanted kidney was rejected, killing nearly every patient within weeks. The one exception, Gregorio Woloshin, was a South American physician whose new kidney somehow worked for five months before it failed.

Moore, impervious to the harsh criticisms from other physicians, took encouragement from Woloshin's case. When Hume left that year for active duty in the Navy, Moore persuaded another young surgeon, Joseph Murray, to continue the work. Murray, a plastic and burn surgeon, shared Moore's omnivorous scientific curiosity and, in his quiet and devoutly Catholic way, the same unflagging optimism. Murray transplanted kidneys into six more patients, saying a prayer for each one. Again, none survived. Fifteen patients died before the doctors accepted that there was no surgical way around the natural process of rejection.

In late 1954, the team received for treatment a twenty-two-year-old named Richard Herrick, who was deathly ill with kidney failure from scarlet fever. He was incoherent and shaken by frequent convulsions; his family was desperate. He was, in other words, like scores of patients the team had seen, except for one detail: he had an identical twin, Ronald. There were a few scattered reports suggesting that skin grafts between identical twins didn't suffer rejection. So how about a kidney transplant? Richard would be the test case. By any measure, it would be a dangerous undertaking. No one could be sure that this transplant would go any better than the others. And now the doctors would be putting a healthy young man at risk—subjecting him to both a major operation and the uncertain long-term consequences of having just one kidney.

Moore approved the operation without hesitation. A series of tests confirmed that the twins were identical. Adding to the pressure, a reporter who was at a police station when the twins arrived for one of these tests, a fingerprint examination, found out about the impending operation and made the news public. On December 23, 1954, under heavy press scrutiny, the urologist J. Hartwell Harrison opened up Ronald and removed one kidney. Moore carried it to the operating room next door, where Murray placed it in Richard's pelvis and sewed its ureter to his bladder and its artery and vein to the nearby blood vessels.

Ronald recovered with ease. More shakily, Richard, too, got better. The donated kidney survived, and there were no signs of rejection. Six months later, Richard returned home from the hospital (and soon married the nurse who had cared for him). This was the first genuinely successful transplant operation in history, and the news spread around the world.

It was, in truth, a freakish kind of victory—how many identical twins needed kidney transplants, after all? But it inspired Moore and Murray and their colleagues to persist. The next year, Moore, operating on dogs, attempted the first liver-transplantation operations ever performed. Murray and the nephrologist John Merrill began experimenting with ways to suppress the rejection response, which, it was now clear, was their primary barrier. The only known immune suppressant at that time was radiation—a lesson learned from the atom bombing of Japan. So, beginning in 1958, they subjected patients to massive doses of radiation—enough to destroy all their antibody-producing cells—before giving them a kidney transplant.

After the irradiation, each patient—even sicker now, with essentially no immune system—was moved into an operating room for its antiseptic environment and one-on-one, around-the-clock medical attention. This was, it turned out, one of the first intensive-care units in the world (and a few years later Moore was among the first to convert an entire ward to function in this way). Many of the patients lived in these converted ORs for weeks. Once stabilized, they were injected with bone-marrow cells from a kidney donor to try to create an immune system that would tolerate donor tissue—the first attempted bone-marrow transplants—and then underwent the kidney transplant itself.

What followed, however, was a series of grisly deaths. The irradiation caused the first patient's blood to lose its ability to clot, and she died from bleeding a few days after the operation. Another patient, a twelve-year-old boy, died from the irradiation without even making it to the operating table. "Mortality was virtually complete," the surgeon Nicholas Tilney recounts in *Transplant: From Myth to Reality,* his history of transplantation. "As a result, the medical, surgical, and nursing staff became increasingly doubtful about the entire enterprise. . . . Indeed, one senior medical resident in charge of the ward finally refused to involve himself any longer, telling John Merrill that he had officiated at enough murders and could not continue." In all, fourteen patients died before the irradiation approach was abandoned, in 1960.

In 1959, researchers at Tufts University Medical School had discovered a drug, 6-mercaptopurine, that could suppress the immune system in rabbits to the point that they could tolerate small injections of foreign cells. Almost immediately, a British surgeon, Roy Calne, began trying the drug in dog kidney transplants. All the animals died, but he reported one that lived forty-seven days without rejection, and Moore persuaded Calne to join his laboratory. Even before Calne arrived in Boston, Moore gave approval for Murray to perform human kidney transplants using the new drug. The first patient was operated on in April, 1960, and died after four weeks. The second died after thirteen weeks.

Murray and Calne tried a modified version of the drug, called azathioprine. Ninety per cent of the dogs they tried it on died within weeks. Still, in April, 1961, the team went forward with a human transplant using the new drug. The patient died after five weeks. The team tried again with another patient, using a different dose regimen. That patient died, too. In the next few months, the surgeons tried four more human kidney transplants using the drug. All the patients died.

Then, in April, 1962, after all those years and all those deaths, the surgeons had a startling success. Murray took a kidney from a man who had died undergoing a heart operation and put it into Melvin Doucette, a twenty-four-year-old accountant. Doucette was immediately, and permanently, put on azathioprine. He underwent two rejection crises, which the doctors managed with steroid injections. And then Doucette went home. He proved to be the first patient in the world to successfully receive a transplanted organ from an unrelated donor. Within months, the Brigham team had performed transplant operations on twenty-seven patients, nine of whom survived long-term. On May 3, 1963, *Time* put Moore, not quite fifty years old, on its cover. Murray went on to receive the Nobel Prize. Their results continued to improve. And kidney transplantation became a routine, life-saving operation around the world.

WHAT KIND OF PERSON can do this again and again—inflict suffering because of an unproved idea, a mere scientific hope? Inflicting suffering is part of any doctor's life, especially a surgeon's. We open, amputate, slice, and burn. I have put needles into screaming children who required two nurses to hold them down. I have put three-foot rubber tubes through people's noses down to their stomachs, though it made them gag and vomit and curse me in several languages. In many cases, I've found the prospect so loathsome that I looked for every excuse to avoid it. But you push ahead, because what you are doing will help the patient. You have experience and textbooks full of evidence that assure you it does. And, because it does help the patients, they forgive you; once in a while, they even thank you.

You also do it because you are part of a little community of people for whom doing such things is normal. I imagine that there's a limit to what I could tolerate doing to another person, no matter how good the evidence might be that it will help. I tell myself, for example, that I could never have been one of those Civil War surgeons who sawed limbs off unanesthetized soldiers to save their lives. Then again, I've circumcised babies without anesthesia just because that's the way everyone else does it.

Moore, however, pushed ahead not only despite suffering and death but also despite slim evidence, repeated failure, and the outright opposition of his peers. He was clearly a maverick. Renée Fox, a sociologist, and Judith Swazey, a historian, have published many studies of the murky, violent territory of surgical innovation, and they found that almost all of the surgeons involved—pioneers in heart transplantation, liver transplantation, the artificial heart—had what they called, in the title of their brilliant 1974 book, "The Courage to Fail."

It is a generous phrase. After all, the surgeons' lives were not the ones on the line. And the portrait that emerges from Fox and Swazey's work is not entirely appealing. These surgeons were ruthless, bombastic, and unbending; they saw critics as either weak or stupid. They were self-absorbed and brutally competitive. They killed patients. They were also stunningly capable leaders: their ability to find the positive in even the worst disasters was infectious. And they hewed to a simple moral stance. Moore used to tell the story of Cotton Mather and Zabdiel Boylston. In 1721, during a deadly smallpox epidemic in Boston, Mather, a clergyman, and Boylston, a doctor's son with no medical training, inoculated two hundred and forty-seven healthy people, including Boylston's own son, with pus containing wild smallpox virus. On the mere basis of an account that Mather had heard from his slave about African practices, they argued that the procedure would be protective. Boston physicians were outraged. But the inoculation worked, and became a standard medical practice. To a physician like Moore, when people were going to die, the moral position was to do something, anything, however dangerous or unproved.

I asked Fox and Swazey how the surgeons they'd come to know coped with the pain and misery and disappointment to which they'd subjected so many patients. The answer varied, the researchers said. For many surgeons, the science itself—the process of reporting successes and failures in papers and at meetings, challenging critics, coming up with modifications when things haven't worked—was how they got by. Other surgeons had to steel themselves to proceed: as they scrubbed in for another try with an unproved operation, they made morbid, self-mocking jokes about the pointlessness of it all, the torture they were about to put some poor fellow through. A few spent hours with their patients and experienced tremendous anguish about what they were doing to them. One notoriously tough pioneer admitted to sleepless nights and long bouts of melancholy. You never knew "at the end of the day, or the end of the decade, or the end of the third of the century, whether what you were striving for was actually going to be anything resembling what you'd hoped it would be . . . whether what was being pawned off as treatment might, in a very real sense, be a disease in and of itself," he told the researchers. The work, he said, "was life-destroying." It was precisely their closeness to their patients that drove such surgeons to proceed. People they knew were dying and desperate, and these surgeons felt that they were the only ones in the world willing to take a chance to help them. On the other end of the spectrum were the pioneers who simply refused to interact with their patients as people. "I don't worry about them after the surgery," one famous surgeon told Swazey. "I just leave all that crap to the chaplains."

Moore was somewhere in between. Nobody saw Moore anguish about the plight of his patients, and he didn't go to their funerals. But he did not avoid his patients, either, however badly things turned out. All his life, he had an odd mixture of prickliness and charm. His enthusiasm and articulate intelligence accounted for the charm, but he was also icy in his convictions; Moore wanted to be right more than he wanted to be loved. He was the sort of man who insisted that his family have breakfast together every morning, despite a schedule that got him to work by six, the sort of man who tolerated no opposition when he'd decided that an experimental treatment should be tried.

Renée Fox pointed out something else about the surgeons she had known: a surprising number of them came to have second thoughts about what they did. Thomas Starzl, for example, was the first surgeon to attempt human liver transplants and also, many deaths later, to succeed with them. He went on to pioneer multiple-organ transplantation, though it took him years to bring the mortality rates down to acceptable levels. Then, later in his career, his courage, or bloody-mindedness, abandoned him. In a recently published interview with Fox, he told of becoming unable to take kidneys from living donors anymore. He had never had a death doing so; but he'd had to amputate the foot of one donor, a professional softball player, owing to complications, and the case haunted him. Over time, he became unable to do experimental work even on dogs.

Moore would never admit to having regrets, or doubts, not even during the most difficult years of kidney transplantation. Yet, finally, something shifted in him, too.

IN 1963, both Moore and Starzl attempted liver transplants, in nine patients. All of them died horribly—bleeding and jaundiced and in shock from liver failure. The researchers called a temporary halt to the procedure and returned to the lab. A year later, Starzl decided to resume the operations, as did Roy Calne, who was now back in London. Instead of joining them, however, Moore denounced them, arguing that survival rates in animals had not improved enough. He never returned to the procedure or allowed others at his hospital to do so. In 1968, after Christiaan Barnard, in South Africa, performed the first successful heart-transplant operation and others around the world raced to follow him, Moore opposed the attempts as premature. He voiced similar concerns about the first efforts at implanting artificial hearts.

"Does the presence of a dying patient justify the doctor's taking any conceivable step regardless of its degree of hopelessness?" he wrote in an influential 1972 statement. "The answer to this question must be negative. . . . It raises false hopes for the patient and his family, it calls into discredit all of biomedical science, and it gives the impression that physicians and surgeons are adventurers rather than circumspect persons seeking to help the suffering and dying by the use of hopeful measures." Cautious experiments in animals were mandatory, he argued, before such measures could be tried in human beings.

His position was hard to fault. Only twelve of the first hundred and thirty liver recipients were long-term survivors. Heart transplantation had even worse results: of the first hundred heart recipients, ninety-eight died in less than six months. Cardiac surgeons were forced to call a moratorium on the procedure, and years passed before animal experiments were thought successful enough to justify broad-scale human trials again. Moore's doctrine became widely accepted policy. New technologies and operations for the terminally ill now require far greater levels of proof in animal trials before use in human beings is accepted. A few years ago, human gene-therapy experiments were brought to a halt by a single death. The doctrine made Moore a hated figure among animal-rights activists—they staked out his house, prompting him to buy a .38-calibre Smith & Wesson for protection. It would also have forbidden nearly every experiment he and the surgeons under him had conducted in the previous thirty years. Moore had helped make medicine bold, unafraid, and at times disturbingly invasive; this was suddenly the era of intensive care, of chemotherapy and open-heart surgery and organ transplantation. But Moore himself never experimented with a major new therapy in human beings again.

Moore changed, as did the rest of us. We no longer tolerate surgeons who proceed as he had once proceeded: we call them cowboys if we're being generous and monsters when someone dies. We prefer the later Moore, the Moore whose almost blind scientific optimism was replaced by caution and carefully modulated skepticism. This was the Moore who was among the few to testify before Congress, in 1971, questioning whether Nixon's War on Cancer had any chance of a rapid victory. As a medical adviser to NASA, starting in the Apollo era, he raised serious, still unanswered concerns about the radiation risks that astronauts might incur from prolonged travel in deep space. Through the nineteen-seventies, he conducted a landmark series of studies of American surgical care and reported unexpectedly high rates of failure and complications.

This was also the humanist who, in the nineteen-sixties, wrote a handbook entitled *The Dignity of the Patient* and required all his surgical residents to read it. He was solicitous in the care of his terminal patients. Although his bedside manner wasn't especially warm—he remained more fatherly than familiar to patients—he expected his residents to regard all the details of patients' care as their responsibility: not just medication doses and test results but the easily forgotten basics of care, such as keeping weak and failing bodies clean, lotioned, neatly dressed, turned and repositioned, free of pain. It was among his most fiercely held criticism of today's medicine that nurses no longer comforted or touched patients the way they used to. And as he became older he grew angry at the possibility that he, too, could die this way—subjected to the kind of routine medical aggressions that his own work had made possible.

MOORE WAS HIMSELF a terrible patient. At a surgical conference in San Antonio, in 1994, he contracted Legionnaires' disease. He was shaking with chills and sick with pneumonia. But he was convinced that the hospital was more likely to kill him than the bacteria were. He hounded the doctors mercilessly, refusing a CT scan that they ordered and demanding that they justify even the antibiotic they prescribed.

When I was a junior resident, I cared for him during a one-night stay after he'd had a hernia repaired. He was eighty years old and questioned the doctors relentlessly about the anatomical details of his hernia, and about whether the repair was done properly. (The surgery failed, as it happened, and he had to return to get his hernia fixed again.)

He was used to thinking of himself as more or less invincible—not an unusual sentiment among doctors, but one that was bolstered by the luck he thought he'd had. During the Korean War, when he was thirty-eight, he was a surgical consultant to the MASH units. Flying along enemy lines, the Army L-5 two-seater he was in came under artillery fire and he was hit. Yet he somehow escaped critical injury, taking a bullet in his right arm, nowhere else, and without any lasting effect. When he was in his seventies, he developed abdominal pains, and a CT scan revealed a large mass in his pancreas. His doctors suspected advanced cancer, yet the mass proved to be benign. Surviving that, he only became surer that he was somehow different from everyone else.

Moore had been physically active all his life, his emphysema notwithstanding. He rode horses, hunted, fished; he acted in, and played music for, society theatre. He sailed competitively, once coming in second in the Halifax Race from Marblehead to Nova Scotia, and he kept up sailing for years after he retired from his surgical practice, in 1981, at the age of sixty-eight. He published scientific papers well into his eighties. He took up a campaign against managed care. In 1997, he and John H. McArthur, the former dean of Harvard Business School, published a landmark article proposing a national council on medical care, modelled on the Securities and Exchange Commission, to oversee the medical and financial standards of commercial health insurers. In 1988, when his wife, Laurie, to whom he had been married for fifty-three years, died in a car crash, he grieved and recovered and then got married again, to Katharyn Saltonstall, the widow of a longtime friend.

Then, in 2000, when he was eighty-seven, his family noticed that he was experiencing shortness of breath. He denied it, of course, but he got winded just walking along the sidewalk. He had to stop every few steps to make it up a flight of stairs. He couldn't continue with his regular sails on Buzzards Bay. Finally, he agreed to see his internist. Listening to Moore's heart, the physician heard a soft *whoosh* where there should have been a *lub-dub*. A weak valve was causing his heart to fail.

Once Moore accepted the diagnosis, he decided that he needed an operation. He wanted a surgeon to open his chest, put his hands inside, and make the problem go away. But his case wasn't bad enough for surgery. The doctors sent him home with prescriptions for drugs he'd have to take for the rest of his life—Lasix (a diuretic, to reduce the fluid congesting his heart and lungs) and Digoxin (a cardiac stimulant). He hated the idea of being dependent on the pills, though, and he didn't take them. His condition worsened.

One day, he couldn't catch his breath. He was admitted to the hospital, where he was given oxygen, threaded with catheters, and infused with enormous doses of diuretics that squeezed gallons of fluid from his lungs. His breathing soon stabilized and, over the next few days, became easier again. Before long, the doctors gave him a handful of prescriptions, and a lecture about taking them, and sent him home.

Thanks to the medical care he had helped create, this eighty-seven-year-old man could have expected to live five and a half more years, according to the National Center for Health Statistics. A gradual, protracted downward spiral is now taken as the norm; it was not a prospect that Moore could regard with any equanimity. His mind was still strong. But he was too frail to travel to scientific meetings. His life had become constricted; he knew he would never feel strong again.

Moore was contradictory on the subject of death. In his later years, he emphasized its inevitability and worried that science was keeping people alive too long. In a memoir, he told the story of an eighty-five-year-old woman he had taken care of who was badly burned in a fire. Her face and upper chest were gone. She was on a ventilator, and would need numerous skin grafts and weeks of care. Her odds of surviving were poor. If she did make it, she'd never eat normally or see normally again. Her husband had died. Her daughter hadn't seen her for years. A grandson came to visit just once. So when her condition began to worsen, Moore went ahead with his own decision. "We began to back off on her treatment," he wrote, and "when she complained of pain, we gave her plenty of morphine. A great plenty. By the clock. Soon, she died quietly and not in pain." He even advocated that doctors recognize euthanasia as a part of their responsibilities.

At other times, however, he insisted that death be fought, forestalled by any means necessary. In 1998, the wife of a close friend had become sick with diabetes, heart disease, and an ischemic, unremittingly painful leg. A vascular surgeon had attempted an arterial bypass to restore blood flow to the leg, but to no avail. Moore sent his son Chip, who was also a surgeon at my hospital, to see her. "There's a syringe of morphine in the drawer," she told Chip. "End this for me." He argued with her. Her vascular surgeon believed that her leg would gradually get better, he said. "He's wrong," she said. Her family needs her, he said. "I have nothing more to give them," she replied. He refused to administer the lethal shot. But when he saw his father he told him that, in his heart, he thought that she was right. The elder Moore exploded in rage that his son would even contemplate it.

ON NOVEMBER 24TH, 2001, two days after Thanksgiving dinner with his son's family, and just after finishing breakfast with his wife, Katharyn, Francis Moore went to his study, shut the door, took out his old Smith & Wesson from the desk drawer, put it in his mouth, and shot himself. In his journal, which his family has kept private, he had evidently contemplated the idea on several occasions. In one entry, a few days earlier, he had written something to the effect that "today is the day," only to follow it with an entry saying that things hadn't seemed right after all. There was nothing self-pitying in what he wrote or said during his last few months. But he had kept the gun clean and hidden for no other purpose than to end his life.

"Was I surprised by his suicide? Not at all," Robert Bartlett, a Michigan surgeon he was close to, told me. Several others said the same thing. I found it unnerving to talk to Moore's surviving friends and colleagues about his decision. "If you ask any of us, we all have some kind of plan," one told me over tea and cookies in his small, neat assisted-living apartment with a red emergency button by the door. He had, he confided, a special bottle of pills in his bathroom cabinet.

There was principle in Francis Moore's death, but also something like anger. He left no note, despite having left a written legacy in every other respect. He shot himself with his wife in the house—to hear it, to find him with his brains sprayed all over the study. There was also, it must be said, a characteristic degree of obliviousness of the pain he could and did cause others.

WHICH MOORE WAS the better Moore? Strangely, I find that it is the young Moore I miss—the one who would do anything to save those who were thought beyond saving. Right now, in the intensive-care unit one floor above where I am writing, I have a patient who is going to die. She is eighty-five years old, a grandmother. She was healthy, on almost no medicines besides a daily aspirin, and lived on her own without difficulty. A few days ago, however, she developed an ache in her abdomen which grew until she was rigid with pain. In the emergency room, we discovered that she had a strangulated hernia, and, in the operating room, we saw that what had been strangulated was a long loop of now purple and dead intestine. We excised the gangrenous bowel, stapled the two ends back together, and fixed her hernia. In the ICU, she improved rapidly. But the gangrene had unleashed an inflammatory response that swept through her whole body. In a process known as the adult respiratory distress syndrome (ARDS), her lungs became stiff, fibrous, and increasingly incapable of exchanging oxygen. Now that we have given her as much oxygen and support as we could and still failed to restore the O_2 levels in her blood to normal, her death seems almost certain.

I spoke to her family outside the sliding glass doors of her room and told them this. They were subdued and unhappy. "Isn't there anything more you can do?" one of her children asked. And what I told him was "No." Strictly speaking, though, that wasn't true. Certainly, we have done everything that the textbooks and journals say we should do. But in an earlier era Francis Moore would have insisted that we use our science and ingenuity to come up with something else. We could try, for example, using a potent recombinant protein called drotrecogin alfa, which was recently found to interfere with inflammation and to reduce mortality in patients suffering from systemic infections. We could try putting her in a hyperbaric oxygen chamber. We could try liquid ventilation. We could try injecting—I read about this once—fluid from pig lungs into her airways. None of these therapies have been shown to work in animals, let alone in adults with ARDS. The expense would be high, and they carry serious risks of increasing her suffering or killing her outright. The recombinant protein, for instance, costs more than ten thousand dollars and can cause severe bleeding. Colleagues and superiors would think I was mad and cruel if I used any of them. But it's not inconceivable that one of them could work, maybe even allow her to go home a week from now.

The family told me that she was clear about her wishes: if there was nothing more we could do to improve her situation, she did not want to be kept on life support. So we will do nothing more. Tonight her heart will slow, then stop completely. She will go peacefully, and that is good. She has had a long and happy life. But I can't help thinking about the costs of our caution. There would once have been a time, under Francis Moore, when we'd have been trying something, anything—and maybe even discovering something new.

JOHN UPDIKE

Mars as Bright as Venus

FROM *THE NEW YORK TIMES BOOK REVIEW*

O brown star burning in the east,
elliptic orbits bring you close;
as close as this no eye has seen
since sixty thousand years ago.

Men saw, but did not understand,
the sky a depthless spatter then;
goddess of love and god of war
were inklings in the gut for them.

Small dry red planet, when you loom
again, this world will be much changed:
our loves and wars, at rest, as one,
and all our atoms rearranged.

K. C. COLE

Fun with Physics

FROM *THE NEW YORKER*

Janet Conrad fell in love with the universe at 3 AM on a cold autumn night in Wooster, Ohio. A teenager, she had no desire to get out of bed and face the frigid air in order to help her father, a dairy scientist, spray warm water on the prize dahlias they were growing together. But when she did go out to the garden she saw, for the first time in her life, how a shower of electrically charged particles flung from a star ninety-three million miles away can cover the sky in glowing pastel curtains. "I remember standing there and looking at the northern lights, and it was so neat that something so remote, so very far away, could be creating something so beautiful right in front of my eyes," she says. Twenty-five years later, Conrad, thirty-nine and an associate professor of physics at Columbia University, created her own universe—a spherical particle detector, forty feet in diameter, that she built under an igloo of dirt at the Fermi National Accelerator Laboratory (Fermilab), near Chicago. The particle detector is lined with a constellation of twelve hundred eight-inch-wide "eyes," or phototubes, and is filled with eight hundred tons of baby oil, which is used to detect the shock waves generated by particle interactions. In the fall of 2002, the detector began an unblinking vigil for subatomic stealth particles known as "sterile" neutrinos.

A lot can go wrong in large-scale physics experiments. Conrad has been basted in foul-smelling oil. She has been squirted with sticky insulating goo.

She has had giant helium balloons get away because the soccer nets she was using to hold them down came loose. And she watched in dismay as the pristine white tank for her current experiment acquired a tough yellow scum (which her mother helpfully advised her to remove with Arm & Hammer baking soda). Nothing that Conrad has done in the past, however, approaches the challenge of her current experiment, which involves some fifty scientists from twelve institutions—including the experiment's co-leader, Bill Louis, of the Los Alamos National Laboratory. Conrad's goal is to understand the character of neutrinos, mere wisps of matter that are more numerous, more elusive, and arguably more important than any other subatomic particle.

Neutrinos outnumber all ordinary particles by a billion to one—a thousand trillion of them occupy your body at every second, streaming down from the sky, up from the ground, and even from radioactive atoms inside you. But, for all their omnipresence, they might just as well be ghosts. As John Updike put it in his poem "Cosmic Gall":

> The earth is just a silly ball
> To them, through which they simply pass,
> Like dustmaids down a drafty hall.

Neutrinos can slip through a hundred light-years' worth of lead without stirring up so much as a breeze, and yet they power the most violent events in the universe, making stars shine and, in the process, creating every element, every dust mote, every raindrop, and, ultimately, every thought. They are the alchemists of the cosmos, the catalysts that make nuclear fusion possible, releasing the radiation that melts rock and makes the continents move. Because neutrinos can penetrate almost everything, they can take scientists to places they've never been before—into the cores of exploding stars, for instance, or back to the big bang. As the universe evolved, neutrinos, because they interact so rarely with other particles, were, in effect, left behind, frozen in time. They are still there (or here, if you will) today, imprinted with information about the state of the universe at its birth. Most important, neutrinos break the basic rules that govern other particles, thereby suggesting that the rules themselves are wrong. If Conrad's experiment confirms her suspicions, she will show that a particle that was barely believed to exist can carry enough weight to determine the drape of galaxies.

When I met up with Conrad at a gathering of the group of Columbia professors who work on high-energy physics, she was the only woman there. Some of her fellow-physicists seemed not to know what to make of her. In contrast to earlier generations of women physicists, she has managed to remain unabashedly girlish. She uses words like "neat" and "cool," and her talks are often embellished with whimsical drawings and analogies to hair dye, shopping, or flowers. "She gets away with it because she knows her stuff so well—nobody can attack her," Bonnie Fleming, a Fermilab physicist, says. Rocky Kolb, a cosmologist at Fermilab, explains, "She obtains what she wants in a different way than most physicists. She can tell you you're wrong without telling you you're stupid. That's unusual. This is not a field for the faint of heart—it's like herding cats. You have a hundred physicists and deep down inside they all think they're smarter than you are. Everyone else herds with a cattle prod. Janet does it with charm." Conrad's charisma has carried her to the top of a field that has traditionally had few places for women, and has won her acclaim and honors such as the prestigious Presidential Early Career Award and the New York City Mayor's Award for Excellence in Science and Technology. (During the presentation of the latter, she had to teach then Mayor Giuliani how to pronounce "quantumchromodynamics.")

Next to her colleagues in khaki and oxford blue, Conrad, with shiny auburn hair, a short cranberry-colored skirt, and matching heels, offered a study in contrasts. She laughs a lot, and her speech comes in staccato bursts, with few transitions. "Conversations seem to jump around, but, if you probe, there's a logical connection," her mentor, the Columbia physicist Michael Shaevitz, says. "She has this wealth of information stored away in different areas, and she pulls it in like an octopus with tentacles." Conrad also gets angry easily. "I have a tendency to fight bitterly," she admits. "I'm so hotheaded that half the time I get myself into trouble." But, she adds, "A little bit of hotheadedness doesn't hurt you in this field." Conrad applied for tenure at Columbia at the first possible moment (and got it, at the age of thirty-six), because she simply couldn't stand the suspense of waiting. A lot of her energy seems to be fuelled by Diet Coke; empty cans line up wherever she goes, like bread crumbs marking her trail. (She drinks so much of it that her mother bought her stock in the company.) It is this combination of charm and restlessness—as well as her ability to design clever and elegant experiments—that has made Conrad successful enough to persuade Columbia to put up a million dollars to get her experiment going while she waited for grant money to arrive. It was an enormous sum for any young physicist to receive, let alone a woman.

NEUTRINOS HAVE BEEN eluding physicists ever since Wolfgang Pauli first hypothesized their existence, in 1930. In the physical universe, what goes in always equals what comes out, in one form or another. But physicists had noticed that when radioactive atoms spat out electrons and transformed into other kinds of atoms, some of the original energy appeared to be missing. Pauli proposed that it had been carried away by a virtually invisible particle. The thought was so preposterous, however, that even he seemed disinclined to take it seriously. "I have hit upon a desperate remedy," he wrote to his colleagues. "But I don't feel secure enough to publish anything about this idea." He went on to express his embarrassment at his own heresy: "I have done a terrible thing. I have postulated a particle that cannot be detected." In 1931, the physicist Enrico Fermi baptized the hypothetical particle "neutrino," or "little neutral one," but his paper was rejected by the journal *Nature* as too "speculative" and "remote from reality."

The first experiment actually to hunt for neutrinos was called, appropriately enough, Project Poltergeist. In 1956, the Los Alamos–based physicists Clyde Cowan and Fred Reines found a definite trace of neutrinos in the intense wash of radiation spewed forth from newly commissioned nuclear reactors. They wrote to Pauli, who reportedly shared a case of champagne with friends. But detecting neutrinos from a reactor was one thing, and detecting them in nature was another. The first neutrinos from the sun weren't discovered until twelve years later, in the Homestake gold mine, in South Dakota, where they created reactions in a tank filled with chlorinated cleaning fluid. To everyone's surprise, however, the Homestake experiment, led by Raymond Davis, Jr., also discovered that about two-thirds of the neutrinos that had been expected to arrive from the sun as a product of nuclear fusion were missing. A 1992 experiment designed to detect atmospheric neutrinos found a similar portion absent. Physicists came up with various theories to account for these disappearances, but in the past few years they have settled on one: neutrinos aren't really "missing"; they are simply altering themselves en route. "You start out with a race of house cats, and you find you have lions in the end," Conrad explains; the physicists set their traps for kittens, and lions ignore the bait.

Of course, it is misleading to think of fundamental particles as clearly defined entities like cats. They are more like waves. You can imagine a particle, for example, as the sound wave you make by plucking a guitar string. A neutrino, however, is not a single defined wave. It is a mixture of waves—messy yet fundamental, like a signature. Neutrinos are known to exist in three different forms, each associated with a member of the electron family—the electron neutrino, the muon neutrino, and the tau neutrino. Yet any neutrino can be part electron, part muon, and part tau neutrino, and it can change as its waves fall in and out of step with each other. The discovery that neutrinos oscillate between forms has huge consequences. The particles were originally thought to be weightless, but in order for the waves to fall in and out of synch they must have some mass. (Guitar strings with different masses produce different sounds because they vibrate at different frequencies; if neutrinos had no mass, there could be no oscillation.) Given their numbers, this means that neutrinos—even if their mass is minute—must weigh as much as all the stars in the sky.

Although physicists can easily tell how many neutrinos go "missing," they can't always tell what forms they have changed into. Most are transforming between the familiar electron, muon, and tau forms. But a highly controversial experiment, conducted at the Los Alamos National Laboratory in 1995, uncovered a fourth possibility: that a portion of the neutrinos were changing into a completely unknown species, not electron, muon, or tau, and perhaps substantially more massive than any of those—a species that was undetectable by any means except its gravitational pull. This hypothetical form is known as the sterile neutrino.

Most physicists were skeptical of the results, in large part because the numbers seemed out of line with those from previous experiments. And, when Conrad and Michael Shaevitz decided to design an experiment to prove or disprove the Los Alamos results, their colleagues were dumbfounded. They couldn't believe, according to Shaevitz, that such "well-respected physicists" would even bother. Conrad's friends asked her why she wanted to "waste her life." In 1998, however, combining her skills as an experimenter and a salesperson, Conrad presented her proposal to the advisory council at Fermilab. In contrast to the many larger experiments performed at the lab, hers was designed to give clear results quickly, relatively inexpensively, and with mostly recycled equipment. Even the project's name was chosen to stress its streamlined approach: Mini Booster Neutrino Experiment (or MiniBooNE). "The name was just weird enough that everyone remembered it," Conrad says—which is no small matter when it comes to securing funding. It was the only time that anyone remembers applause at such a meeting. "We usually sit there and scowl," Andreas Kronfeld, a Fermilab physicist, says. "Then this sparkling young person gives a really good talk. It was a little sunshine in all that gray."

FERMILAB WAS FOUNDED in 1967 by Robert Rathbun Wilson, a Berkeley-trained physicist who had worked on the Manhattan Project and then, after the bombing of Nagasaki, refused to take part in weapons research. Wilson was also a sculptor, and he planted his works everywhere on the sixty-eight-hundred-acre grounds. There is a Möbius strip in a pool on the roof of the auditorium, a staircase modelled on a double strand of DNA. Even the utility poles are shaped like the symbol for pi. Thousands of physicists come here now, mostly to conduct experiments with the world's highest-energy accelerator, a circular racetrack for particles called the Tevatron. Around its four-mile circumference, superconducting magnets steer protons travelling at almost the speed of light into head-on collisions, setting off fireworks of particles as the energy of speed congeals into matter. In effect, each collision creates a miniature big bang—creation all over again, thousands of times per second. Shopping-mall-size detectors (think of them as elaborate electronic eyes) are needed to keep track of just a tiny part of this activity.

The Tevatron ring is bordered, on the outside, by a river of cooling water, like a moat. Inside, Wilson and his successors have reverentially re-created twelve hundred acres of Illinois prairie. If Fermilab was to be the frontier of physics, Wilson reasoned, then it should have its roots in the literal frontier. The grasses grow up to ten feet tall, and much of the original wildlife has returned, including deer, foxes, salamanders, turtles, beavers, weasels, mink, and hundreds of species of birds. As a final touch, Wilson added a herd of buffalo. There is another poignancy at Fermilab these days, a palpable sense that billion-dollar particle accelerators and physics labs—the "cathedrals of contemporary science," as Wilson called them—have fallen from favor since the post–Second World War period, when physics produced not only the H-bomb but also such leaders in the disarmament effort as Einstein and Wilson himself. In many ways, this makes the results of small experiments such as MiniBooNE even more pivotal.

I attended a MiniBooNE meeting at Fermilab a few months before the experiment was due to be launched. The meeting took place in a glass-walled high-rise building that Wilson modelled on Beauvais Cathedral, in France. It was a difficult time for the project. The head contractor had been killed in a car accident the day before the meeting, and a building designed to house part of the experiment was still unfinished. A shipment of baby oil from Exxon had turned out to be unusable, and the company was struggling to replace it. "A million things that I never thought could happen have happened on this experiment," Conrad said. "You'd never dream that your contractor could die." Many of the researchers had gathered to bring one another up to date on the experiment's progress: the people who would generate the beam of muon neutrinos, which Conrad and her collaborators hoped to observe oscillating into "sterile" forms; the people building the detector; the people in charge of the oil and of the computer programs to keep track of it all; and the people working on the physics itself—the properties of neutrinos, and possible astrophysics applications. There was conversation about dump-chunks, QTsmear, spillsplitters, and Roefitters. Someone passed around a heavy ring, a custom-made flange. It was an odd pairing of mathematics and metal, of the ephemeral and the concrete. "That's one of the neat things about being an experimentalist," Conrad says. "You can actually see what you've accomplished."

CHILDHOOD IN A SMALL Ohio town was, it turns out, a surprisingly good preparation for a career in particle physics. Conrad learned many of her practical skills in the local 4-H Club. "Electronics really isn't that different from cooking or sewing," she says. "There's a certain set of rules that you follow, a certain set of patterns. You may want to try variations on the theme, but, once you know your patterns, it's pretty easy." Conrad believes that she was probably born to be a scientist. "I loved having conversations with my parents about how the world works," she says. "They never treated me as someone who couldn't understand." Her uncle, Walter Lipscomb, a Nobel laureate in chemistry, challenged her whenever he came to visit, throwing out puzzles at the dinner table. "I was a miniature adult as far as he was concerned, and he was happy to let me in on his world," Conrad says. After her sophomore year at Swarthmore, her uncle offered her his apartment in Cambridge, Massachusetts, for the summer, and suggested some people she might approach for jobs at Harvard. One of them was the late physicist Frank Pipkin, who was using the Harvard cyclotron to test parts of an experiment to be installed at Fermilab. Conrad worked with Pipkin that summer. The following summer, she went with him to Fermilab. As soon as she saw the detectors, she says, "I knew I wanted to play with them. It was big and dirty and it was just so me."

Dirt, it seems, is an important ingredient in particle-physics experiments. The otherwise flat topology of Fermilab is interrupted by big mounds of earth, long ridges that look as if some large, determined animal were burrowing beneath. The earth acts as an insulator, protecting the experiments below from stray cosmic rays, and the people above from radiation produced by the particle beams. The tank of baby oil that is the heart of MiniBooNE sits under a dirt hill that Conrad describes as "remarkably like a home for Teletubbies." Bill Louis, Conrad's partner, drove me to the site. We bounced over unpaved tracks until we came to something that looked like a large yurt with prairie grasses sprouting from the top of it. A small antechamber inside was stacked with metal cabinets full of computers. Louis opened a square entry hatch and extended a rickety metal ladder, and we crawled down backward into the huge white tank, which was about to be sealed for good.

Building such an experiment can involve some difficult choices. The quarry you're chasing can't be seen directly, so you have to induce it to leave visible traces. The world's most famous neutrino experiment, Super-Kamiokande (or Super-K), in Japan, uses fifty thousand tons of purified water—watched over by twelve thousand phototubes—to catch the wakes produced by the traceable particles in neutrino interactions. Workers ride inside the vast cavern in a little rubber boat. MiniBooNE uses oil, which leaves slightly better wakes but precludes the possibility of boat travel. ("If I fell out, I would sink to the bottom and die," Conrad says, laughing.) The wake is actually a shock wave made by an electrically charged particle travelling through the oil—something like a sonic boom—and the phototubes pick it up as a ring of light. A fuzzy ring signals that the incoming neutrino was an electron neutrino; a sharp one signals a muon neutrino. But it's not always that simple. Other kinds of reactions may be impossible to identify. "Until you've tried to work with a particle as elusive as a neutrino, you have no idea how hard this can actually be," Conrad says. There are so many decisions to be made: Do you watch for neutrinos to disappear or for puzzling appearances? And where, exactly, do you look? Say you start off with lions and they change into cats after two miles, then back into lions two miles later; if you placed your detector at the four-mile mark, you'd see only lions and could easily conclude that nothing had happened. "So often you pose a question, and you build a detector to look at it," Conrad says. "Then the detector answers some other question."

Conrad took me to a hangarlike area, where a critical part of the experiment was being tested. It was an enormous aluminum "horn" designed to focus pions, unstable particles that naturally disintegrate into neutrinos as they travel through a hundred and fifty feet of sewer pipe buried underground. A purified beam of neutrinos would then continue on, through ordinary ground, toward the detector full of oil. To anyone who has spent time in a physics lab, this seems peculiar: other sorts of particle beams are steered through pristine vacuum pipes to insure that they aren't bumped off their course by unintended collisions with air. Neutrinos, however, don't "see" air, any more than they see lead; everything is a vacuum to them. Conrad expects neutrinos from the beam to collide with molecules in the oil several times a minute. But, for every controlled neutrino encounter, a hundred thousand will be caused by cosmic rays from the atmosphere. To avoid these false signals, most neutrino experiments are performed at the bottom of deep mines. MiniBooNE, instead, will receive its beam of neutrinos in bursts, five every second. By comparing the timing of the bursts and of the signals, Conrad and her collaborators hope to sort the needles from the hay. Three hundred extra phototubes line an outer wall of the MiniBooNE sphere, forming a "veto region" designed specifically to flag stray signals—a muon entering from the outside, say, rather than being created by a neutrino in the oil.

IN CONRAD'S TENTH-FLOOR OFFICE at Fermilab, a field of fake gerbera daisies swayed on tall metal rods like orange and red lollipops. The bottom shelf of a bookcase held a small bottle of Tide detergent and a black light (to demonstrate how scintillators glow), tuning forks (for explaining oscillating neutrinos), and little vials of oil. Conrad never passes up an opportunity to discuss what she's doing. On planes, she draws other passengers into conversations, inviting them to visit Fermilab. She recently completed a radio series, *Earth and Sky,* which aired on National Public Radio. And she works with high-school teachers, and even students, whom she hires to help her on experiments. When she won the Maria Goeppert-Mayer Award—for outstanding achievement by a young woman physicist—in 2001, she began to turn her attention more directly to the problem of bringing women into her field. She acknowledges that women tend to have a different style of doing physics from men; they use more words relative to equations, for instance, and this can count against them on exams. "You can watch the guys look at this and say, 'Too many words. Must not know what she's talking about,' " Conrad says. But, in response to those who question whether the specialized equipment is intimidating to women, she insists that it's a lot less complicated than what you find in an average well-stocked kitchen. Conrad's presence has already made a big difference in the Columbia physics department. When she first started teaching there, eight years ago, a site report by the American Physical Society concluded starkly, "Columbia is not a friendly place for women students and perhaps for students in general." By 2000, half of the undergraduates in the high-energy-physics program were female, a shift that many people attribute to Conrad's aggressive attempts to draw young women into the department.

When colleagues at Columbia talk about what she has accomplished, they often tell the story about the benches. In the fall of 1996, she noticed that students were sitting on the dirty floor in the hallway, waiting for her class to start. When she asked for benches to be installed, the administration balked at the expense. Over the Thanksgiving break, she was visiting a former student's family and mentioned the problem. The next day, the student's father, a Columbia alumnus, mailed her a check for a thousand dollars, which she used to buy five plain benches. Other people in the department liked the look of them, and started to add their own touches to the hallway—posters and plants. Then, as Conrad tells it, "the university people came back and said, 'Wow, this looks really nice, but these benches are not the topnotch benches we would like to have here.' So they went and bought nice benches and gave me back mine." "It's one of those things that really made a difference," Steve Kahn, then the chair of the physics department, says.

AFTER OUR DAY at Fermilab, Conrad took me to her house in nearby Geneva, Illinois. She has two other homes as well—an apartment in Manhattan and a house in Las Cruces, New Mexico, where her husband, Vassili Papavassiliou, teaches physics at New Mexico State University. Conrad and Papavassiliou met while they were wiring two tiers of the same experiment (one is reminded of Lady and the Tramp sharing that fateful strand of spaghetti), but, like many academic couples, they couldn't find work in the same city. They make do, often meeting at Chicago's O'Hare airport for dinner. "It's really bad when the waitress at the airport starts to know you," Conrad said.

A suitcase full of dirty laundry had waited by the door of Conrad's Geneva house for several days, but the night before she had prepared two Greek pies for our dinner, brushing thirty separate layers of phyllo dough with butter—after a full day's work and a friend's fortieth-birthday party. "That's why people don't like to be around me much," she said. "I wear them out." The previous weekend, she had planted three hundred tulip and daffodil bulbs. Spread out on the coffee table were photographs of her father's dahlias, which she was examining in order to choose roots for next year. She called them "bursts of light," and they did seem to explode, like red and pink and orange fireworks against the black backgrounds. "I look at these beautifully symmetric flowers, and yet I find that the one little flaw in them is the thing that makes the flower interesting," she said. "And that's true in high-energy physics, too. It's the little flaws that make it fascinating. And some of the little flaws are not so little."

Of course, it's hard to know whether a flaw is just a flaw, or whether it's a crack in the edifice of physics, a first glimpse into something entirely unexpected. Because data are always ambiguous, it can be years before physicists feel confident enough to publish potentially controversial results. On September 1, 2002, neutrinos began to trickle into the baby oil at MiniBooNE. By mid-June, 2003, the detector's phototubes had already picked up a hundred thousand interactions. Conrad says that she won't know for at least two years whether the disparaged Los Alamos results were right after all. But it should be worth the wait. If MiniBooNE eventually proves the existence of the sterile neutrino, it will require physicists to rethink everything, from the details of the big bang to the formation of the elements. It could even help explain why there is matter in the universe at all. On the other hand, it will be just as important if Mini-BooNE proves that there is no sterile neutrino. "Not finding the ether was a successful experiment," Rocky Kolb says, referring to an experiment that proved there was no medium for carrying light—a finding that helped to cement Einstein's theory of relativity. "A lot of people have too much ego to work on confirming experiments," he adds. "Janet is one of the world's leaders in neutrino physics, and, whatever the future of particle physics is, she'll play a major role."

As MiniBooNE began to take in data, a second experiment gearing up at Fermilab was preparing to send a beam of neutrinos four hundred and fifty miles to a deep mine in Minnesota. Another experiment aims to send neutrinos from Geneva, Switzerland, through the Alps to a lab under a mountain near Rome. An endeavor aptly named ICECUBE will turn a cubic kilometre of ice at the South Pole into a detector to observe neutrinos from the stars and cores of galaxies. Conrad, meanwhile, has taken a lead role in pushing for a National Underground Science Laboratory in the Homestake mine, where neutrinos from the sun were first observed to be missing. (Raymond Davis, who led the first Homestake discovery, won the 2002 Nobel Prize in Physics.) And in 2002, Conrad also completed her work on a select panel of physicists charged with deciding what kind of major particle accelerator should follow Fermilab's designated successor, now under construction in Europe. As usual, Conrad was one of the youngest people on the panel, and there were some difficult moments. Some people blamed her for orchestrating a protest against a few of the panel's recommendations, even though she claims she had nothing to do with it. "I believe I make things clot," she says. "You add me to a mixture, and all of a sudden big chunks of stuff will fall to the bottom."

Two days after I left Fermilab, one of the twelve thousand phototubes in Super-K, the Japanese detector, collapsed while it was being refilled with water after routine maintenance. The shock wave from the collapse created a storm inside the tank. Seven thousand phototubes were shattered. "This is a real disaster," Conrad told me over lunch in Madison, Wisconsin. She'd driven up from Fermilab with Len Bugel, a physics teacher at Stratton Mountain School, in Vermont, with whom she'd worked for many years, and in the car she had persuaded Bugel to try to figure out how to duplicate the effect on a small scale—to see whether MiniBooNE's tubes would collapse under the same conditions. "Creative resource-getting is the No. 1 thing you have to learn," Conrad said. And, by the way, she asked, was I interested in spending the summer at Fermilab, helping out on her experiment? "We'll teach you what you need to know," she said.

OLIVER MORTON

Strange Nuggets

FROM *WIRED*

[Editor's Note: Scientists propose to track that most elusive and yet most plentiful of physical phenomena, "dark matter," by using the entire planet Earth as their detector.]

On the morning of October 22, 1993, local time, the visitor hit the top of the atmosphere over Ellsworth Land in Antarctica. It pierced the sky in a flash of light, moving a hundred times faster than a meteor, passing from the thinnest air and into the ice in a fraction of a second. It cut through the rock below with equal ease, flying through the solid earth in a northeasterly direction. In less than 20 seconds, it had crossed the South Atlantic, deep beneath the ocean floor. When it passed below the southern tip of Africa, it was more or less halfway between Cape Town and the center of Earth.

That was as deep as its straight path through the planet would take it; from then on it headed up. Fifteen seconds later, 6,000 kilometers across the Indian Ocean from Cape Town, it left Earth's crust somewhere between Sri Lanka and Thailand. It lanced up through the afternoon sky and headed back out to the stars. The whole visit lasted less than a minute, and nobody saw a thing.

It's remarkable that some strange guest should sweep through Earth like a hot wire through wax, and that no one would notice as it did so. But though the visitor was very fast and fairly heavy, it was also extremely small: a mass of as much as 10 tons squeezed into something about the size of a red blood cell. If a 10-ton asteroid fell to Earth at 400 kilometers per second, people would notice; something the size of a small car hitting the unyielding Earth at that speed would give up its kinetic energy in an explosion to rival that of a 200-kiloton nuclear weapon. But condensed to the size of a small amoeba, the same mass wouldn't cause anywhere near as much fuss. The fearsome momentum of the microscopic visitor would shatter the bonds between molecules directly in its path and push the bystanders aside. It would do this vigorously enough to melt a small tunnel as it passed, slicing through the rocky earth almost as easily as it passed through air and water.

If there had been a human at the point of entry, she might have seen a split-second burst of light, and the exit through the ocean must have made a fish-startling noise of some sort. But no one saw or heard a thing. No one, that is, for almost 10 years. It was only then that scientists noticed what might have been the visitor's faint signals in an obscure archive of seismological data. Such data piles up at ever-increasing rates as Earth gets more thoroughly wired; every passing planetary groan and tremor now gets recorded, preserved to be pondered by anyone who has the time, training, and curiosity to look. And in the past few years, a small team of physicists has developed the expertise to trawl that data for signs of intruders from outer space. These are researchers who, when the grandest physics project ever conceived was canned, set themselves a task far smaller in cost but arguably grander in ambition: surreptitiously turning a whole planet into a particle detector.

VIC TEPLITZ MET Eugene Herrin in 1989. Both were on the faculty of the physics department at Southern Methodist University in Dallas. It was a heady time to be a particle physicist in Texas. Excavation was about to start on the 87 kilometers of underground tunnel, located an hour's drive from campus, that would house the Superconducting Super Collider, the Next Big Thing in science's endeavor to understand the makeup of the universe. The SSC, it was hoped, would produce all sorts of wonderful particles for the world's physicists to study, like the much sought-after Higgs boson. "For a few years," recalls Teplitz, "we were the capital of the planet."

Unfortunately for Texas, funding for the construction of the Superconducting Super Collider was canceled in 1993, and hundreds of people like Teplitz suddenly found a collider-shaped hole in their careers. One response was to turn from particle physics to particle astrophysics. The universe is full of things that can produce energies far greater than the SSC would ever have managed. And there's evidence that space may be richly endowed with bizarre particles forged in the almost inconceivable energies of the big bang itself. The behavior of the bits of universe that astronomers can see—the spinning of galaxies, for example—convinces them that there must be a lot of stuff out there they can't see, invisible remnants of creation that outweigh all the visible stars and galaxies. Since the demise of the Texas collider, attempts to identify this "dark matter," or at least to observe its effects, have become something of a boom industry.

Most of the dark-matter candidates that researchers find interesting are extraordinarily light and tiny. But there are exceptions. One of these is "strange matter," a substance a bit denser than the nuclei of atoms. Normal atomic nuclei are made of particles called up-quarks and down-quarks—two ups and a down to make a proton, two downs and an up to make a neutron. The heavier quarks created in accelerators—the strange quark, the charmed quark, the bottom quark, the top quark—are normally unstable. But it's possible that, under certain circumstances, strange quarks could be stabilized, if mixed in with everyday up-quarks and down-quarks. While large atomic nuclei are unstable, lumps of strange matter would hold together fine at almost any size. A piece of strange matter could be as massive as a star and still not fall apart.

So, what would it mean for Earth if the dark matter that astronomers believe envelops our galaxy was made of strange matter? Strange nuggets up to a billion or so times the mass of a normal atom would fall to Earth and just sit there, chemically inert and hard to find. Larger nuggets would penetrate the planet's interior before stopping. And nuggets weighing more than a tenth of a gram would pass right through. A large nugget, elbowing its way through Earth at high speed, might be detectable by seismologists.

After the cancellation of the SSC, Herrin and Teplitz decided to pursue this idea—not least because Southern Methodist had some excellent seismologists. Suspecting that seismic signals given off by strange nuggets might be detectable with current technology, they decided to sift through a suitable data set—a stack of 10 big old reel-to-reel tapes from the US Geological Survey, containing more than 9 million seismic events reported by stations around the world between 1981 and 1993. Slowly, with the help of David Anderson, a computer expert at SMU, and Ileana Tibuleac of Weston Geophysical in Boston, the scientists started to make sense of the data. This wasn't glamorous work; no atoms were smashed, no deep space probed. It meant dealing with dusty data, gathered in small rooms in out-of-the-way places by people watching cathode rays wiggle across their screens, or inky traces squiggle along rolls of paper. Put all that information together, though, and Earth becomes a sounding board—a far larger, far less precise analog to the fine-tuned machines that record the passing of particles produced by machines like the SSC.

The reports the team studied were of "unassociated events," vibrations that had set seismometers trembling enough for human overseers to take note, but that had not subsequently been ascribed to a particular earthquake with a known epicenter. For each year, there were about a quarter-million such reports. From among those, the SMU team screened out any reports of events taking place within an hour of a quake for which an epicenter had been calculated by the USGS. For 1993, that left 152,272 reports. Then they picked out the reports from the most sensitive "Class I" seismic stations. "There are 5,000 stations at various stages of operation, and it would have been prohibitive to look at them all," says Herrin. "So we chose the most sensitive ones and those that most consistently reported to the geological survey." That left 38,866 reports. Next, they looked for any seven or more that were clustered within 20 minutes of one another—since with fewer than seven measurements, it wouldn't be possible to calculate the source's trajectory. Once they'd winnowed down the data, they ran an algorithm to see if a linear source—one that moved through Earth in a straight line—could account for the seven signals.

Earthquakes, which are seismologists' sustenance, appear to release their energy from a single point in the crust; so do nuclear explosions, which are Herrin's specialty. As a result, seismologists have plenty of algorithms for tracing tremors from a wide area back to a single point source. But scientists had never developed an algorithm that would find linear sources, for the simple reason that they didn't imagine there were any.

The first linear source found by the algorithm was the October 22, 1993, event. Between 09:58:52 Universal Time—when a signal was picked up at Gauribidanur, India—and 10:06:30—when a signal faded away at La Paz, Bolivia—there were reports from two stations in Australia, one in Turkey, and two others in Bolivia. For six of the seven reports, the signal's arrival time was within a second of that predicted for a linear source beginning in Antarctica and passing under the southern tip of Africa to the northeastern Indian Ocean. The two Australian stations had the wherewithal to calculate the direction of the source; they showed the signals to have come from that part of the hypothetical track closest to Australia—just as the team predicted. When Herrin and Teplitz looked at stations in Africa and New Zealand, which might have been expected to pick something up from the hypothetical linear source, they saw nothing—but the noise levels at the stations were high enough that they would not have expected to see anything. When they tried to see whether there was a traditional point-source model that could explain the data, the algorithms failed to find anything.

IF THE OCTOBER 22 CLUSTER was good, the cluster a month later was even better. On November 24, the algorithm spotted a set of nine reports from Australia and South America between 10:26 and 10:29 UT.

A linear source starting south of French Polynesia and ending 20 seconds later in the Ross Sea of Antarctica, quite close to McMurdo Station, could explain all the signals to within less than half a second, a much better result than you'd get for most run-of-the mill earthquakes.

The timing of the signals was not the only striking thing; their shapes were odd, too. "The waveforms for the two events are extremely unusual," says Herrin. "They're prolonged—energy continues to come in for a significant length of time compared to what you see from small earthquakes, where you usually have a little sharp burst of energy." The Australian signals came in for 30 seconds, the La Paz signal a bit longer. No single point-source model can account for that.

Even better was the news that the Australian signals had a sudden drop in amplitude after a few seconds. That drop corresponds to the time the nugget would have left Earth; the fainter signals afterward are from farther back along the nugget's trans-Pacific path, arriving in its wake, just as the sound of a supersonic jet arrives only after the aircraft has passed. Says Herrin, "We think that sudden drop is very distinctive."

THE TEXAS TEAM'S RESULTS have been on the Web since summer 2002, and the reaction to them has been instructive. Seismologists who look at the analysis find it very intriguing and want to know more. The fact that Herrin is well respected as a "reasonably cautious man," in the words of Ray Willemann, the director of the International Seismological Center in southern England, is undoubtedly helping get the work taken seriously.

But if those who make a living listening to Earth are intrigued, the cosmologists and particle physicists who think about dark matter are more reticent.

Part of the problem is that they would really rather not have their dark matter made up of strange nuggets. Particle physicists are interested in dark matter because it might be made of particles they haven't yet been given enough money to make in their laboratories. If the dark matter is made up of dull old quarks, it's less exciting (and less likely to inspire new funding). What's more, to appreciate the strength of the Herrin team's data, you need to think about various aspects of geophysics, which cosmologists and particle physicists—the discipline's stars and aristocrats—don't often deign to do. The general response from the particle people and the cosmos crowd is a noncommittal grunt of "come back when you've got more data."

Some of this may explain why the journal *Nature* has rejected the paper Herrin and his colleagues submitted on the subject (it's now being reviewed by the considerably lower-profile *Bulletin of the Seismological Society of America*). Meanwhile, scrutiny by other seismologists has revealed some data that complicates the story: Records from the International Seismological Center show a small earthquake in the Indian Ocean about five minutes before the first of the Texas group's events. This doesn't necessarily mean the team's linear trace is a figment but does make the visiting-nugget formulation a little harder to nail down. The November 1993 event, though, still looks good.

The team's biggest problem, however, is not the indifference of astrophysicists or the difficulties of publication. It's that they have a result they can't reproduce. Seismologists may have unwittingly turned Earth into a great big strange-nugget detector—but they haven't yet wired things up so that the detector is easy to read: The unassociated-events database the Texas team used stops in late 1993. By that time, most of the best seismic stations had moved over to digital data recording, which made it easier to archive continuous streams of data rather than reports people made as the data came in. It may make more sense to record raw data than the preliminary analysis of what that data means. But for people looking for strange events, the old system's humans-in-the-loop selectivity had been a useful filter. Without the unassociated-events file to start from, the scientists would have to trawl through dauntingly huge databases, seismometer by seismometer, looking for long, drawn-out anomalous readings; cross-correlate all the findings; and then search for linear sources.

The ideal solution would be to give up on the databases and look at worldwide reports in real time. Of particular interest to Herrin and Teplitz is the data from the International Monitoring System being set up to verify the Comprehensive Nuclear Test Ban Treaty. The IMS will be, when finished, a truly superb way of monitoring Earth. Its 50 primary seismological stations, backed up by existing arrays around the world, are meant to pick up any nuclear test down to 1 kiloton; and it's expected the system will be even more sensitive than that. With that degree of sensitivity and global coverage, the IMS would be a useful readout tool for a planet-sized strange-nugget detector.

What's more, the IMS has other components that might be able to help. It includes an infrasound system for picking up long-wavelength atmospheric noises and a hydrophone system for listening to the oceans. Both are superbly receptive: The infrasound instruments allow scientists in Holland to keep track of Mount Etna's eruptions and researchers in Los Alamos to detect meteorite fireballs over the central Pacific. The hydrophones can pick up noises from thousands of kilometers away—researchers think it might be possible to conduct cetacean censuses by listening to the sum of the ocean's whale song. Either of these systems might, in principle, be able to catch the distinctive sound of a nugget on its way in or out of the earth.

That sound—whatever it might be—would be powerful confirming evidence for the nuggets hypothesis. As Teplitz points out, all the team has shown so far is that there appear to be linear events *within* Earth. Evidence of something heading in or out would go a long way toward proving that it's really cosmic in origin.

Unfortunately, as yet there is no way of looking at the data from the growing International Monitoring System in real time. While the United States, which is paying for a substantial fraction of the IMS, is eager to see all the data released in near-real time, other parties to the Test Ban Treaty, most notably China, are keen to keep the data confidential. The treaty itself, according to Oliver Meier, a verification expert at the London-based organization Vertic, is not clear on this point. What's more, the US position is undermined by the fact that it now opposes the treaty.

"We're looking at what kind of computer capability is required for different sorts of searches," says Herrin, "but we're definitely going to go forward with the research." He's particularly interested in the possibility that some linear events may have been recorded originally as earthquakes at the point of exit. "Our assumption that all events are point sources rules all the algorithms. But if even one of these events is accepted as linear, then all the science of seismology is going to have to deal with it." At the same time, he and Teplitz are looking at how much of the energy the nuggets give off gets turned into seismic signals; Herrin thinks the proportion is higher than that for a nuclear explosion, which wastes a lot of energy cracking nearby rocks, and thus that the trackable nuggets may be lighter than first thought—a ton, say, rather than 10 tons.

ONE TON OR 10, though, the nuggets would still be too big to be of much interest to fundamental physicists, who are more interested in the properties of single particles they haven't yet observed than of vast agglomerations of particles they already know of. Indeed, that is pretty much Teplitz's point of view, too. His personal tradeoffs aside (asked if strange nuggets would be a fair replacement for what could have been achieved at the SSC, he says, "In discovering the Higgs, I'd have had a minor role; in looking for strange nuggets in seismic records, our team is alone"), Teplitz is clear that the Higgs and its kin are important in a different way: They reveal things about how the universe works, rather than just add to the list of wonders it contains.

If not fundamental, though, quark nuggets zipping around the galaxy would still be an amazing addition. And perhaps even more amazing, in the end, than any technically strange—or just generally bizarre—particles burrowing through the ground would be the fact that the planet is no longer just a block of dumb rock in their path. It is an ever better wired planet, monitored and thought about in ever more ingenious ways; it is a datasphere ever more sensitive to its surroundings and its own processes, from flashes in the upper atmosphere to rumblings in the core. We have made it a planet that notices things. We have made it an observant Earth.

KEAY DAVIDSON

Mapping of Cosmos Backs Big Bang Theory

FROM THE *SAN FRANCISCO CHRONICLE*

NASA astronomers have unveiled history's best-ever "baby picture" of the universe—the most accurate, detailed snapshot of the cosmos close to the beginning of time.

The stunning achievement helps answer many questions that have taunted cosmologists, including the age and ultimate fate of the universe.

The image, recorded by a NASA satellite in deep space some 1 million miles from Earth and announced Tuesday, February 11, 2003, at a press conference, shows the cosmic microwave radiation, the foggy afterglow that pervaded the cosmos 380,000 years after the Big Bang.

Ripples in the radiation mark primordial building blocks of "superclusters"—immense clouds, chains and sheets of galaxies that crisscross today's universe like gossamer superhighways, National Aeronautics and Space Administration scientists announced.

By analyzing the ripples, scientists have concluded the cosmos is 13.7 billion years old; that it will expand forever until it dissipates like a cloud; that it consists mostly of mysterious "dark matter" and "dark energy"; and that the first stars began forming much sooner than originally thought.

NASA called the achievement by the Wilkinson Microwave Anisotropy

Probe, or WMAP, "one of the most important scientific results of recent years." Its data "lay the foundation for a unified and coherent cosmic theory," said WMAP principal investigator Charles Bennett at the press conference.

The WMAP satellite data is helping scientists answer many key questions:

- Three decades ago, astronomers debated whether the universe had always existed or began with the Big Bang. The new pictures confirm the reality and certain details of the Big Bang and reveal the age of the cosmos as 13.7 billion years, roughly three times as old as Earth itself.
- To their surprise, scientists have concluded based on WMAP images that the earliest stars formed only 200 million years after the Big Bang. That's several hundred million years earlier than originally thought, NASA said at the press conference at Goddard Space Flight Center in Greenbelt, MD.
- Thanks to the WMAP images, scientists conclude that only 4 percent of the cosmos consists of matter as we know it. The rest of the universe is so-called "dark matter" and "dark energy," about which virtually nothing is known.

"We live in an implausible, crazy universe," said one of the grand old men of American cosmology, astrophysicist John Bahcall of the Institute for Advanced Study in Princeton, NJ. He wasn't a member of the WMAP team but participated in the NASA press conference.

"Every astronomer will remember the moment he heard the results from WMAP," Bahcall said. He acknowledged that he had previously questioned the idea that the universe is pervaded by an invisible "dark energy" that is causing the cosmos to expand faster with time. But the WMAP data has changed his mind.

WMAP is a joint project of NASA and Princeton University. It is named after David Wilkinson, a Princeton cosmologist and WMAP researcher who died in September 2002. Other team members came from UCLA, the University of Chicago, Brown University, and the University of British Columbia in Vancouver.

In June 2001, NASA launched the 16-foot-long, 1,800-pound WMAP to its present location a million miles from Earth, four times the distance to the moon.

Over 12 months, WMAP's sensors scanned the entire sky. They recorded the extremely faint "cosmic background radiation" or CBR, which is microwave radiation that pours from all directions of the sky. First detected in the mid-1960s, CBR is the afterglow of the Big Bang.

During those 12 months, WMAP was, in effect, taking a yearlong "time exposure" of the CBR. The feat is "the equivalent of taking a picture of an 80-year-old person on the day of their birth," says a NASA summary of WMAP findings.

Most important is WMAP's mapping of "ripples" in the CBR with unprecedented precision.

When first discovered in the 1960s, the CBR displayed seemingly uniform intensity in all directions of the sky. That uniformity puzzled astronomers. It appeared too "smooth" to explain how gravity could have tugged together matter fast enough to form galaxies and superclusters of galaxies since the Big Bang.

The mystery began to clear up in 1992, when another satellite, NASA's Cosmic Background Explorer, or COBE, revealed subtle nonuniformities, or ripples, in the background radiation.

WMAP's observations provide a far higher-resolution "map" of the cosmic background radiation. So high, in fact, that it reveals ghostly "structures" as small as a few hundred thousand light-years across, roughly the size of the larger galaxies. (A light-year is 6 trillion miles, the distance light travels in a year.)

NEIL deGRASSE TYSON

Gravity in Reverse

FROM *NATURAL HISTORY*

Sung to the tune of "The Times They Are A-Changin' ":

> Come gather 'round, math phobes,
> Wherever you roam
> And admit that the cosmos
> Around you has grown
> And accept it that soon
> You won't know what's worth knowin'
> Until Einstein to you
> Becomes clearer.
> So you'd better start listenin'
> Or you'll drift cold and lone
> For the cosmos is weird, gettin' weirder.

> —*The Editors (with apologies to Bob Dylan)*

Cosmology has always been weird. Worlds resting on the backs of turtles, matter and energy coming into existence out of much less than thin air. And now, just when you'd gotten familiar, if not really comfortable, with the idea of a big bang, along comes something new to worry about. A mysterious and universal pressure pervades all of space and acts against the cosmic gravity that has tried to drag the universe back together ever since the big bang. On top of that, "negative gravity" has forced the expansion of the universe to accelerate exponentially, and cosmic gravity is losing the tug-of-war.

For these and similarly mind-warping ideas in twentieth-century physics, just blame Albert Einstein.

Einstein hardly ever set foot in the laboratory; he didn't test phenomena or use elaborate equipment. He was a theorist who perfected the "thought experiment," in which you engage nature through your imagination, inventing a situation or a model and then working out the consequences of some physical principle.

If—as was the case for Einstein—a physicist's model is intended to represent the entire universe, then manipulating the model should be tantamount to manipulating the universe itself. Observers and experimentalists can then go out and look for the phenomena predicted by that model. If the model is flawed, or if the theorists make a mistake in their calculations, the observers will detect a mismatch between the model's predictions and the way things happen in the real universe. That's the first cue to try again, either by adjusting the old model or by creating a new one.

One of the most powerful and far-reaching theoretical models ever devised is Einstein's theory of general relativity, published in 1916 as "The Foundation of the General Theory of Relativity" and refined in 1917 in "Cosmological Considerations in the General Theory of Relativity." Together, the papers outline the relevant mathematical details of how everything in the universe moves under the influence of gravity. Every few years, laboratory scientists devise ever more precise experiments to test the theory, only to extend the envelope of its accuracy.

Most scientific models are only half baked, and have some wiggle room for the adjustment of parameters to fit the known universe. In the heliocentric universe conceived by the sixteenth-century astronomer Nicolaus Copernicus, for example, planets orbited the Sun in perfect circles. The orbit-the-Sun part was correct, but the perfect-circle part turned out to be a bit off. Making the orbits elliptical made the Copernican system more accurate.

Yet, in the case of Einstein's relativity, the founding principles of the entire theory require that everything take place exactly as predicted. Einstein had, in effect, built a house of cards, with only two or three simple postulates holding up the entire structure. (Indeed, on learning of a 1931 book titled *100 Authors Against Einstein*, he responded, "Why one hundred? If I am incorrect, one would have been enough.")

That unassailable structure—the fact that the theory is fully baked—is the source of one of the most fascinating blunders in the history of science. Einstein's 1917 refinement of his equations of gravity included a new term—denoted by the Greek letter lambda—in which his model universe neither expands nor contracts. Because lambda served to oppose gravity within Einstein's model, it could keep the universe in balance, resisting gravity's natural tendency to pull the whole cosmos into one giant mass. Einstein's universe was indeed balanced, but, as the Russian physicist Alexsandr Friedmann showed mathematically in 1922, it was in a precarious state—like a ball at the top of a hill, ready to roll down in one direction or another at the slightest provocation. Moreover, giving something a name does not make it real, and Einstein knew of no counterpart in the physical universe to the lambda in his equations.

EINSTEIN'S GENERAL THEORY of relativity—called GR by verbally lazy cognoscenti—radically departed from all previous thinking about the attraction of gravity. Instead of settling for Sir Isaac Newton's view of gravity as "action at a distance" (a conclusion that discomfited Newton himself), GR regards gravity as the response of a mass to the local curvature of space and time caused by some other mass. In other words, concentrations of mass cause distortions—dimples, really—in the fabric of space and time. Those distortions guide the moving masses along straight-line geodesics, which look like the curved trajectories that physicists call orbits. John Archibald Wheeler, a physicist at Princeton University, put it best when he summed up Einstein's concept this way: "Matter tells space how to curve; space tells matter how to move."

In effect, GR accounts for two opposite phenomena: good ol' gravity, such as the attraction between the Earth and a ball thrown into the air or between the Sun and the Earth; and a mysterious, repulsive pressure associated with the vacuum of space-time itself. Acting against gravity, lambda preserved what Einstein and every other physicist of his day had strongly believed in: the status quo of a static universe. Static it was, but stable it was not. And to invoke an unstable condition as the natural state of a physical system violates scientific credo: you cannot assert that the entire universe is a special case that happens to be precariously balanced for eternity. Nothing ever seen, heard, or measured has acted that way in the history of science. Yet, in spite of being deeply uneasy with lambda, Einstein included it in his equations.

Twelve years later, in 1929, the US astronomer Edwin P. Hubble discovered that the universe is not static after all: convincing evidence showed that the more distant a galaxy, the faster that galaxy is receding from the Earth. In other words, the universe is growing. Embarrassed by lambda, and exasperated by having thus blown the chance to predict the expanding universe himself, Einstein discarded lambda, calling its introduction his life's "greatest blunder."

THAT WASN'T THE END of the story, though. Off and on over the decades, theoreticians would exhume lambda—more commonly known as the "cosmological constant"—from the graveyard of discredited theories. Then, sixty-nine years later, in 1998, science exhumed lambda one last time, because now there was evidence to justify it. Early that year two teams of astrophysicists—one led by Saul Perlmutter of Lawrence Berkeley National Laboratory in Berkeley, California; the other by Brian Schmidt of Mount Stromlo and Siding Springs Observatories in Canberra, Australia—made the same remarkable announcement. Dozens of the most distant supernovas ever observed, they said, appeared noticeably dimmer than expected—a disturbing finding, given the well-documented behavior of this species of exploding star. Reconciliation required that either those distant supernovas acted quite differently from their nearer brethren, or else they were as much as 15 percent farther away than the prevailing cosmological models had placed them.

Not only was the cosmos expanding, but a repulsive pressure within the vacuum of space was also causing the expansion to accelerate. Something had to be driving the universe outward at an ever-increasing pace. The only thing that "naturally" accounted for the acceleration was lambda, the cosmological constant. When physicists dusted it off and put it back in Einstein's original equations for general relativity, the state of the universe matched the state of Einstein's equations.

TO AN ASTROPHYSICIST, the supernovas used in Perlmutter's and Schmidt's studies are worth their weight in fusionable nuclei. Each star explodes the same way, igniting a similar amount of fuel, releasing a similarly titanic amount of energy in a similar period of time, and therefore achieving a similar peak luminosity. Hence these exploding stars serve as a kind of yardstick, or "standard candle," for calculating cosmic distances to the galaxies in which they explode, out to the farthest reaches of the universe.

Standard candles simplify calculations immensely: since the supernovas all have the same wattage, the dim ones are far away and the bright ones are nearby. By measuring their brightness (a simple task), you can tell exactly how far away they are from you. If the luminosities of the supernovas were not all the same, brightness alone would not be enough to tell you which of them are far from Earth and which of them are near. A dim one could be a high-wattage bulb far away or a low-wattage bulb close up.

Fine. But there's a second way to measure the distance to galaxies: their speed of recession from our Milky Way, a recession that's part and parcel of the overall cosmic expansion. As Hubble was the first to show, the expansion of the universe makes distant objects race away from us faster than the nearby ones do. By measuring a galaxy's speed of recession (another straightforward task), you can deduce its distance from Earth.

If those two well-tested methods give different distances for the same object, something must be wrong. Either the supernovas are bad standard candles, or our model for the rate of cosmic expansion as measured by galaxy speeds is wrong.

Well, something *was* wrong in 1998. It turned out that the supernovas are splendid standard candles, surviving the careful scrutiny of many skeptical investigators. Astrophysicists were left with a universe that is expanding faster than they had ever thought it was. Distant galaxies turned out to be even farther away than their recession speed had seemed to indicate. And there was no easy way to explain the extra expansion without invoking lambda, the cosmological constant.

Here, then, was the first direct evidence that a repulsive pressure permeated the universe, opposing gravity. That's what resurrected the cosmological constant overnight. And now cosmologists could estimate its numerical value, because they could calculate the effect it was having: the difference between what they had expected the expansion to be and what it actually was.

That value for lambda suddenly signified a physical reality, which now needed a name. "Dark energy" carried the day, suitably capturing our ignorance of its cause. The most accurate measurements done to date have shown dark energy to be the most prominent thing in town.

THE SHAPE OF OUR four-dimensional universe comes from the relation between the amount of matter and energy that inhabits the cosmos and the rate at which the cosmos is expanding. A convenient mathematical measure of that shape is usually written as the uppercase Greek letter omega (Ω). If you take the matter-energy density of the universe, and divide it by the matter-energy density required to just barely halt the expansion (known as the "critical" density), you get omega.

Because both mass and energy cause space-time to warp, or curve, omega effectively gives the shape of the cosmos. If omega is less than one, the actual mass-energy falls below the critical value, and the universe expands forever in every direction for all of time. In that case, the shape of the universe is analogous to the shape of a saddle, in which initially parallel lines diverge. If omega is equal to one, the universe expands forever, but only barely so; in that case the shape is flat, preserving all the geometric rules we all learned in high school about parallel lines. If omega exceeds one, parallel lines converge, and the universe curves back on itself, ultimately recollapsing into the fireball whence it came.

At no time since Hubble discovered the expanding universe has any team of observers ever reliably measured omega to be anywhere close to one. Adding up all the mass and energy they could measure, dark matter included, the biggest values from the best observations topped out at about 0.3. Since that's less than one, as far as observers were concerned, the universe was "open" for the business of expansion, riding a one-way saddle into the future.

MEANWHILE, BEGINNING IN 1979, Alan H. Guth, a physicist at MIT, and others advanced an adjustment to big bang theory that cleared up some nagging problems. In brief, Guth explained why things look about the same everywhere in the universe. A fundamental by-product of this update to the big bang was that it drove omega toward one. Not toward one-half. Not toward two. Certainly not toward a million. Toward one.

Scarcely a theorist in the world had a problem with that requirement, because it helped get the big bang to account for the global properties of the known universe. There was, however, another little problem: the update predicted three times as much mass and energy as observers could find. Undeterred, the theorists said the observers just weren't looking hard enough.

At the end of the tallies, visible matter alone could account for very little of the critical density. How about the mysterious dark matter? Nobody knows what dark matter is, but observers knew there is five times as much dark matter as visible matter. They added that in as well. Alas, still way too little mass-energy. The observers were at a loss. "Guys," they protested, "there's nothing else out there." And the theorists answered, "Just keep looking."

Both camps were sure the other camp was wrong—until the discovery of dark energy. That single component raised the mass-energy density of the universe to the critical level. Yes, if you do the math, the universe holds three times as much dark energy as anything else.

A SKEPTICAL LOT, the community of astrophysicists decided they would feel better about the result if there were some way to corroborate it. The Wilkinson Microwave Anisotropy Probe (WMAP) was just what the doctors ordered and needed. This NASA satellite, launched in 2001, was the latest and best effort to measure and map the cosmic microwave background, the big bang's blueprint for the amount and distribution of matter and energy in the universe. Astrophysicists can now say with confidence that omega is indeed equal to one: the matter-energy density of the universe we know and love is equal to the critical density. The tabulation? The cosmos holds 73 percent dark energy, 23 percent dark matter, and a measly 4 percent ordinary matter, the stuff you and I are made of.

For the first time ever, the theorists and observers kissed and made up. Both, in their own way, were correct. Omega is one, just as the theorists demanded of the universe, even though you can't get there by adding up all the matter—dark or otherwise—as they had naively presumed. There's no more matter running around the cosmos today than had ever been estimated by the observers. Nobody had foreseen the dominating presence of cosmic dark energy, nor had anybody imagined it as the great reconciler of differences.

So WHAT IS this stuff? As with dark matter, nobody knows. The closest anybody has come to a reasonable guess is to presume that dark energy is a quantum effect—whereby the vacuum of space, instead of being empty, actually seethes with particles and their antimatter counterparts. They pop in and out of existence in pairs, and don't last long enough to be measured. Their transient existence is captured in their moniker: virtual particles.

But the remarkable legacy of quantum mechanics—the physics of the small—demands that we give these particles serious attention. Each pair of virtual particles exerts a little bit of outward pressure as it ever so briefly elbows its way into space. Unfortunately, when you estimate the amount of repulsive "vacuum pressure" that arises from the abbreviated lives of virtual particles, the result is more than 10^{120} times bigger than the value of the cosmological constant derived from the supernova measurements and WMAP. That may be the most embarrassing calculation ever made, the biggest mismatch between theory and observation in the history of science.

I'd say astrophysicists remain clueless—but it's not abject cluelessness. Dark energy is not adrift, with nary a theory to call home. It inhabits one of the safest homes we can imagine: Einstein's equations of general relativity. It's lambda. Whatever dark energy turns out to be, we already know how to measure it and how to calculate its effects on the cosmos.

Without a doubt, Einstein's greatest blunder was having declared that lambda was his greatest blunder.

A remarkable feature of lambda and the accelerating universe is that the repulsive force arises from within the vacuum, not from anything material. As the vacuum grows, lambda's influence on the cosmic state of affairs grows with it. All the while, the density of matter and energy diminishes without limit. With greater repulsive pressure comes more vacuum, driving its exponential growth—the endless acceleration of the cosmic expansion.

As a consequence, anything not gravitationally bound to the neighborhood of the Milky Way will move away from us at ever-increasing speed, embedded within the expanding fabric of space-time. Galaxies now visible will disappear beyond an unreachable horizon. In a trillion or so years, anyone alive in our own galaxy may know nothing of other galaxies. Our—or our alien Milky Way brethren's—observable universe will merely comprise a system of nearby stars. Beyond the starry night will lie an endless void, without form: "darkness upon the face of the deep."

Dark energy, a fundamental property of the cosmos, will, in the end, undermine the ability of later generations to comprehend their universe. Unless contemporary astrophysicists across the galaxy keep remarkable records, or bury an awesome time capsule, future astrophysicists will know nothing of external galaxies—the principal form of organization for matter in our cosmos. Dark energy will deny them access to entire chapters from the book of the universe.

Here, then, is my recurring nightmare: Are we, too, missing some basic pieces of the universe that once was? What part of our cosmic saga has been erased? What remains absent from our theories and equations that ought to be there, leaving us groping for answers we may never find?

DENNIS OVERBYE

One Cosmic Question, Too Many Answers

FROM *THE NEW YORK TIMES*

C all it the theory of anything.

Einstein once wondered aloud whether "God had any choice" in creating the universe. It was his fondest hope that the answer was no.

He and subsequent generations of physicists have hoped that at the end of their labors there would be one answer—a so-called Theory of Everything—that would explain why the details of the world are the way they are and cannot be any other way: why there was a Big Bang, the number of dimensions of space-time, the masses of elementary particles.

For 20 years, physicists have lodged those hopes in string theory, a mathematically labyrinthian effort to portray nature as made up of tiny wriggling strings and membranes, rather than pointlike particles or waves.

Once called a piece of 21st-century physics that had fallen into the 20th century by accident, string theory has become one of the hippest fields of science, celebrated in books like the recent best seller *The Elegant Universe*, by the Columbia theorist Brian Greene, and the subject of a miniseries on *Nova*.

In principle, strings can unite all the forces of nature, including gravity, in a single mathematical framework. But the "stringiness" of nature manifests itself only at energies and temperatures that can be generated in a particle accelerator the size of a small galaxy.

As a result, physicists have been left at the mercy of their mathematical imaginations or sifting cosmological data for hints of a clue from God's own particle accelerator, the Big Bang.

The hope was that when all was said and done, there would be only one solution to the theory's tangled equations, one answer corresponding to only one possible universe. But recent progress in string theory paradoxically seems to leave physics further than ever from that dream of a unique answer. Instead of a single answer, the equations of string theory seem to have so many solutions, millions upon millions of them, each describing a logically possible universe, that it may be impossible to tell which one describes our own.

In a series of conceptual and technical breakthroughs, a group of theorists at Stanford showed in 2003 that string theory could describe a universe whose expansion was accelerating—something that many experts thought impossible.

That was no small accomplishment because cosmologists now theorize that some puzzling and so far unidentified "dark energy" is wrenching space apart ever more violently. This energy seems to make up 70 percent of the cosmos, according to astronomical observations.

The new calculations suggest that this dark energy cannot last forever, that it will disappear sometime in the far future, according to the researchers, Dr. Shamit Kachru, Dr. Renata Kallosh and Dr. Andrei Linde, all of Stanford, and Dr. Sandip P. Trivedi of the Tata Institute of Fundamental Research in Bombay.

But the same calculations confirmed that string theory could have a vast number of solutions, each representing a different universe with slightly different laws of physics. The detailed characteristics of any particular one of these universes—the laws that describe the basic forces and particles—might be decided by chance.

As a result, string theorists and cosmologists are confronted with what Dr. Leonard Susskind of Stanford has called "the cosmic landscape," a sort of metarealm of space-times. Contrary to Einstein's hopes, it may be that neither God nor physics chooses among these possibilities, Dr. Susskind contends. Rather it could be life.

Only a fraction of the universes in this metarealm would have the lucky blend of properties suitable for life, Dr. Susskind explained. It should be no surprise that we find ourselves in one of these.

"We live where we can live," he said.

Dr. Susskind conceded that many colleagues who harbor the Einsteinian dream of predicting everything are appalled by that notion that God plays dice with the laws of physics.

Among them is Dr. David Gross, director of the Kavli Institute of Theoretical Physics in Santa Barbara, Calif., who said, "I'm a total Einsteinian with respect to the ultimate goal of science."

Physicists should be able to predict all the parameters of nature, Dr. Gross said, adding, "They're not adjustable."

But Dr. Max Tegmark, a cosmologist at the University of Pennsylvania, said, "I think this grand dream is basically dying."

Dr. Michael Douglas of Rutgers and the Institute of Advanced Scientific Studies, near Paris, called the plethora of string universes "the Alice's Restaurant" problem.

"You can get anything you want at Alice's Restaurant," he said. "Is this a theory of something, very many things or nothing? That's what we're trying to establish."

THE QUESTION OF WHETHER strings will provide a unique answer to the universe has been hanging over physicists' heads ever since the modern form of string theory made its triumphal emergence in 1984. That year, Dr. John Schwarz of the California Institute of Technology and Dr. Michael Green, now of Cambridge University in England, showed that thinking of elementary particles as little strings instead of points eliminated troublesome mathematical anomalies from theories that sought to combine gravity with subatomic physics.

Even Einstein had failed to unite those disparate and mathematically incompatible realms. But the 1984 calculation raised the hope that physicists had finally found the key to the so-called Theory of Everything.

"There was this wild enthusiasm, unbridled enthusiasm, that we paid for later," said Dr. Andrew Strominger, a professor of physics at Harvard.

In 1985, Dr. Strominger, Dr. Edward Witten of the Institute for Advanced Study in Princeton, Dr. Gary T. Horowitz, now at the University of California at Santa Barbara, and Dr. Philip Candelas, now at the University of Texas, published a classic paper showing that it was possible to construct a string theory consistent with the so-called Standard Model that describes particles and forces in our four-dimensional universe.

One problem is that string theory requires 10 dimensions of space-time, whereas we appear to live in four. Dr. Strominger remembered being excited when he found a paper by the mathematician Dr. Shing-Tung Yau, now of Harvard and the Chinese University of Hong Kong. It proved a conjecture by Dr. Eugenio Calabi, now retired from the University of Pennsylvania, that the extra dimensions could be curled up in microscopically invisible ways like the loops in a carpet.

The paper described only one way this folding could be done. But Dr. Yau soon told the physicists that there were thousands of what are now called Calabi-Yau spaces, each one representing a different solution of the string equations. By the time their paper was finished, "the uniqueness of string theory was certainly already in question," Dr. Strominger said.

That was just the beginning. For each of the thousands of ways of curling the extra dimensions into Calabi-Yau spaces, there were hundreds of variations in details like the sizes of the loops and the way electrical and magnetic fields thread through them. When the variations are taken into account, the number of solutions and the number of possible universes can easily exceed 10^{100}.

THIS BOUNTY OF POSSIBILITIES makes it extremely daunting for scientists who want to test string theory by comparing its predictions to the real world. One telltale clue to the right answer, as well as a huge challenge, developed five years ago when astronomers discovered that the expansion of the universe was apparently accelerating. But until recently, many theorists doubted that strings could produce even one example of an accelerating universe.

The reason is that the leading explanation for this behavior is a cosmic repulsion, known as the cosmological constant, that results from the properties of empty space itself.

It was first invented by Einstein in 1917 as a fudge factor to stabilize the universe and then abandoned by him when astronomers found out that the universe was not static, but expanding.

If Einstein's fudge factor is real after all, the universe will continue to expand faster and faster as space grows bigger and bigger, producing more and more repulsion.

String theorists did not know how to deal with the cosmological constant. According to quantum mechanics, the weird laws that govern subatomic physics, empty space should be foaming with energy and particles that wink in and out of existence, and their collective effect could produce a repulsive force like Einstein's constant. But the calculations also suggest that this force should be some 10^{60} times what astronomers have measured; it would have blown the universe apart in its first millisecond, long before atoms, galaxies or humans could form.

Moreover, a permanently accelerating universe would present deep conceptual problems, several physicists pointed out, including Dr. Thomas Banks of Rutgers and the University of California at Santa Cruz, Dr. Willy Fischler of the University of Texas, Dr. Susskind and Dr. Witten.

Such a universe would slowly empty itself of energy and information because most of the galaxies would eventually be flying away so fast that humans could not see them. The observable universe would actually shrink, as if surrounded by a black hole. Life would become impossible, and the usual methods of formulating physics might not apply.

As a result of such arguments, it was widely presumed that a universe that accelerated forever—that is, one with a cosmological constant—was incompatible with string theory, Dr. Kachru of Stanford said.

It was partly to counter such claims, he added, that he and his colleagues were motivated to look for the cosmological constant in the gazillions of possible string universes.

In THE WINTER OF 2003, Dr. Kachru and his colleagues succeeded in using string theory to construct universes that accelerated, but with a surprising twist.

The hitch, in each case, was that the acceleration would be only temporary. It might last an extremely long time, but eventually the dark energy of the cosmological constant would melt away, decaying just in time to avoid the problems of permanent acceleration that string theorists have worried about. The universe would then coast for the rest of eternity.

The work followed on previous work by Dr. Kachru with Dr. Joseph Polchinski of the Santa Barbara Institute and Dr. Steven Giddings of the University of California at Santa Barbara, and by Dr. Polchinski and Dr. Raphael Bousso of the University of California at Berkeley.

Part of the reason dark energy decays, explained Dr. Linde of Stanford, is that these solutions describe the four-dimensional universe we observe around us—three dimensions of space and one of time—with the other six curled up so tightly that they cannot provide closet space. But it takes energy to keep the extra dimensions confined.

"In the long run," he said, "the universe doesn't want to be four-dimensional. It wants to be 10 dimensions."

So sooner or later, the loops will unravel like a tangle of rubber bands, passing through a succession of configurations that take less and less energy to maintain, until finally the other dimensions expand and the cosmological constant is gone.

The decay of the cosmological constant will be fatal, astronomers agree. At that moment a bubble of 10-dimensional space will sweep out at the speed of light, rearranging physics and the prospects of atoms and planets, not to mention biological creatures.

"What it leaves behind," Dr. Susskind said, "it's hard to say. Almost certainly not a livable universe."

THE UNIVERSE IS certainly livable now, but why has long been a vexing and polarizing issue. Life as We Know It seems to require an almost miraculous juggling of a few atomic and astronomical parameters.

Was the universe designed for us? Or did we just get lucky?

Searching for answers, some theorists have invoked the so-called anthropic principle, which states that our universe has to have laws suitable for life. Otherwise we would not be here to see it. The apparent "fine-tuning" of this universe is simply an artifact of our own existence here as observers, Dr. Brandon Carter, now at the Paris Observatory in Meudon, argued in 1974.

The principle fits well with recent theories of the Big Bang that suggest that the universe seen through telescopes is just one in an endless chain of bubble universes that sprout from one another.

If there is just one universe, the fact that it suits us would seem suspiciously lucky. But if there are many universes to choose from, our existence is less miraculous.

It might be the diversity of string-theory universes that gives this metacosmos a chance at harboring life, Dr. Susskind says.

He likes to portray it as a mountain range, the "cosmic landscape," in which the height of the peaks represents the energy or the cosmological constant of that configuration. The universe is like water rolling around the hills, always seeking a lower state. There are valleys and basins and plateaus where it can rest. But it can spread, plopping like a wave sloshing over the hills from valley to valley, from one configuration of dimensions and fields to another.

As a result, he said, in whatever form it starts, the universe will branch out into other forms. If it keeps sloshing, it will eventually land in a valley with the lucky mix of cosmic constants that allows for galaxies and carbon-based chemistry somewhere. If a small fraction of the subuniverses can support life, then there is a good chance that life will arise somewhere, Dr. Susskind explained.

Others caution, however, that it has not been proved that different classes of universes would be so interconnected. "It could be that there are many disconnected landscapes," explained Dr. Douglas of Rutgers.

Dr. Susskind said that "whether we like it or not," the new findings gave further credence to the anthropic principle and a mathematical framework for how it might work.

But such "anthropic thinking" is defeatist to many physicists. "We see this kind of thing happen over and over again as a reaction to difficult problems," Dr. Gross said. "Come up with a grand principle that explains why you're unable to solve the problem."

The notion that some problems are unsolvable is discouraging to the younger generation, he said, pointing out that nobody even knows what the final form of string theory will be.

Dr. Witten said he also disliked the anthropic principle. "I continue to hope that we are overlooking or misunderstanding something and that there is ultimately a more unique answer," he wrote by e-mail.

Dr. Susskind conceded that he had once been on the other side of the question.

"I've had myself jerked around by this theory," he said. "When you have to give up your fondest dream for what the theory would do"—a reference to the quest for a unique answer—"that's a hard thing to swallow."

DR. STROMINGER of Harvard said the debate on anthropic principle was indicative of the "all-or-nothing psychology" of string theory.

"It's not enough to solve some problems," he said. "It has to solve every problem."

Theorists have long hoped that all but one of these solutions will eliminate themselves through some mathematical inconsistency or failure to reproduce an essential feature of the universe like the cosmological constant. Dr. Douglas of Rutgers has challenged that hope, saying string theory may have so many solutions that physical measurements can not distinguish among them.

Indeed, he pointed out in a recent paper, it has not been proved that string theory does not have an infinite number of solutions. So far, anything seems possible.

Rather than sifting myriad solutions for the one that fits our universe, Dr.

Douglas has developed statistical methods to analyze the set of string solutions as a whole to find patterns that will not show up when the solutions are examined one by one.

The results could help ascertain which features of this "zoo of possibilities" are more common and which are more rare, and how many solutions really are too many.

"My own philosophy," Dr. Douglas said in an interview, "is that we should do our best to listen to what string theory is trying to tell us. It is smarter than we are."

Dr. Kachru suggested that it might be wishful thinking to expect that a "smoking gun" confirmation of string theory could be found from comparing it to today's universe. The full glories of string theory, he said, manifest themselves only at energies trillions of times what earthbound particle accelerators can produce.

Perhaps, he said, theorists should be looking for the smoking gun in the Big Bang.

Asked what the smoking-gun question might be, Dr. Kachru laughed and said, "If I knew, I would be working in that field."

SHERWIN B. NULAND

How to Grow Old

FROM *ACUMEN*

I have been thinking a great deal about a long-dead Frenchman named Brown-Séquard, whose distinguished reputation as a medical scientist dissolved nearly overnight when he pridefully reported that he had discovered a treatment to stave off certain ravages of aging, especially those having to do with sexual performance.

Charles Édouard Brown-Séquard was born in 1817 on the island of Mauritius to an American father and a French mother. A brilliant researcher, he made many notable contributions to the understanding of the nervous system and metabolism during his career, and was rewarded by being named professor of experimental physiology at the Collège de France in 1878. He and his predecessor in the Collège chair, Claude Bernard, are properly credited with introducing the notion of hormones, those proteins secreted by ductless glands into the bloodstream, which control so much of the functioning of the internal organs. So impressed was Brown-Séquard by the role of hormones in energizing the animal body and supporting its stability that he began to experiment with them in an effort to rediscover youth.

In 1889, when he was 72 years old, Brown-Séquard reported to the French Academy of Sciences that he had been conducting self experiments in rejuvenation. His method was to crush the testicles of guinea pigs or dogs and inoculate himself with a solution of the fluid thus obtained. Within three days of starting the treatments, he boasted, "I had recovered at least all of my former vigor . . . My digestion and the working of my bowels have improved considerably too . . . I also find mental work easier than I have for years." And, he added, he had regained his sexual prowess.

Unfortunately, inoculating others with the same material had little effect. The treatment may not have helped Brown-Séquard very much either, because he died five years later without demonstrating so much as the most minimal objective evidence that he had accomplished anything in the interim except to age in the usual manner of septuagenarians.

Brown-Séquard's attempt to regain his youth became such an object of derision that it besmirched his scientific heritage. But that did not deter others from involving themselves in similar undertakings, whether with testicular or ovarian extracts or the implantation of the organs themselves. Some of the experimenters were established scientists, but others were hucksters in search of a fast buck. A Kansas charlatan named Charles R. Brinkley became wealthy by implanting goat gonads into suckers, supposedly to treat not only aging and impotence, but high blood pressure as well. No amount of debunking by physicians or the press in the 1920s and '30s could dim his star, which soared to a height so lofty in the firmament that he eventually ran for governor of his state, electioneering from his own radio station. There is not a shred of evidence that any animal's testicles, ovaries, or similar implanted or injected tissue ever helped a single man or woman to return to youth or sexual potency. No one's aging process was halted, and no one became younger.

Brown-Séquard keeps entering my thoughts because of a conversation I had only a few weeks ago with a college classmate whom I had not seen in decades. In the midst of the usual comparisons of our former youthful vigor with the current era of widened waistlines and rusting joints, he casually dropped the news that he was scheduled to receive a penile implant several days hence. A widower of more than a year, he had discovered the tantalizing attractions of much younger women, and he found himself sufficiently unpredictable in sexual performance that he was determined to do something about it. Because we are some 50 years out of college, I wondered aloud how many men of our considerable age had undergone such a procedure. My classmate replied that his urologist had assured him that he was far from being the oldest patient in the hospital's surgical series. His next sentence was what started me thinking about Brown-Séquard. "It's not just the sex itself," he said. "It's the feeling that I'm not giving in to old age." It was at this point that I looked more carefully at my erstwhile pal and realized that he was wearing a toupee.

Many are the ways of "not giving in to old age," and they have ranged from the pitiably ridiculous to the healthily sublime. Brown-Séquard and every hawker of youth-restoring nostrums since has appealed to the universal fear of getting old and to the even greater terror that every new proof of our age brings us further along the road to death. There was general joy a few years ago when molecular biologists began to spread the news that genetic engineering might be the way to stave off the inevitable. But that was a mere prelude to the excitement aroused when telomere research was rumored to hold the promise of preventing cells from developing the degenerating changes that gradually cause them to become senile and finally end their lives. The telomere is a cap like structure at the end of each chromosome that becomes smaller each time the cell divides, thereby seeming to reflect and perhaps influence the number of divisions that are possible. It appears possible, at least to the optimists, that a gene coding for an enzyme called telomerase, which has the ability to maintain or even increase the length of the telomere, may hold the clue to lengthening not only cellular life but the life of the whole organism—even the organism that is you and me.

Upon hearing word of the new miracles seemingly just beyond the horizon, many a middle-aged man or woman—and no doubt plenty of younger ones too—has gone to bed with futuristic sugarplums dancing in their heads. The air is rife with possibilities, among which a life span of two or more centuries is not the most extreme. Too few stop to consider the predictably harmful consequences of such an achievement, not only on human society, but on individuals as well—including the newly spared bicentenarians themselves, cluttering the planet with their needs, their demands, and their refusal to get out of the way. Too few question whether being able to live a very long time is a good thing at all.

The unalterable biological fact is that we, like all animals, exist in order that we may pass our DNA on to succeeding generations. This is how species survive; this is how natural selection works. Once we have lost the ability to do that and perhaps to nurture our young for a while, we serve no useful function in the grand scheme of nature's relentless events. In the wild and even among domesticated animals, death shortly after the end of the reproductive years is the rule. Man is the only animal to survive much beyond that point. But even as late as the Roman Empire, when modern *Homo sapiens* had already been in existence for some 40,000 years, average life expectancy was less than 30 years; infectious disease and inadequate nutrition were the big killers. Two millennia later, in 1900, that figure still had not gone much beyond 45, though most people were far better nourished and protected from contagion. Americans can mostly thank public health measures like clean water, immunizations, improved food supply, and good housing for the current expectation they will reach their late 70s. Though the great advances in biomedicine of recent decades have had an effect too, only antibiotics influence general mortality statistics to a great extent. All the other pharmaceuticals and surgical tinkerings affect relatively small numbers of people in the overall figures for the entire world population.

Aging is not a disease. It is the condition upon which we have been given life. The aging and eventual death of each of us is as important to the ecosystem of our planet as the changing of the seasons. When William Haseltine, PhD, the brilliant biotechnology entrepreneur who is the CEO of Human Genome Sciences, says, "I believe our generation is the first to be able to map a possible route to individual immortality," we should cringe with distaste and even fear, not only at the hubris of such a statement but also at the danger it poses to the very concept of what it means to be human. The current biomedical campaign against the natural process of aging is but part of a much larger conception of humankind's future, in which it is thought by some that parents may one day order up the IQ, complexion, and stature of their intended offspring by manipulating their DNA.

These are not the problems American medicine should be struggling with. Its proper task is not the prolongation of life beyond the naturally decreed maximum span of our species (which seems to be in the neighborhood of 120 years), but its betterment. And if anyone's life needs betterment it is surely the elderly man or woman still living well beyond the years of vigor and productivity because the benisons of public health and biomedicine have made it possible. The percentage of the aged in our population increases with every passing year, and far too many of these people are doddering. The very gradual increase in life expectancy of previous generations has been replaced by a surge forward: the 20th century saw a 33-year gain, an astonishing figure compared to any comparable period in history. Until these recent changes, population size had the general configuration of a pyramid with a wide base of children, the top narrowing with age. It has now taken on somewhat the shape of a rectangle, as more aged individuals reach the upper levels. As disease treatment continues to improve and public health measures reach a larger segment of the population, this trend will only increase.

Some illustrations provide graphic evidence of these patterns. During the 1990s, the number of American centenarians went from 37,000 to 70,000 and is projected to reach a million by the middle of this century. The United Kingdom's Queen Elizabeth customarily sends a telegram to every one of her subjects who attains a hundredth birthday. In the first year of her reign, 1952, the number was 255; it is now well over 5,000 per year. When the United States' Social Security system was established in 1935, it was thought that the system would never have to serve more than 25 million people. There were more than 38 million beneficiaries in 2000, and it is estimated that the ripening of the baby boomer generation will bring the figure to 70 million by 2011.

As much as we might hail such statistics, they quite obviously come with a price. Joints, bones, hearts, brains, and every other part of us lose their zip, and worse. Long-term care institutions are filled with men and women so incapacitated that they require help with the simplest of needs, like toileting and dressing themselves. Many of them are demented. The number of what geriatricians call "the oldest old"—those over 85—increases with each passing year. The economic cost is high, but the cost in suffering, not only for the elderly themselves but for their families, is even greater. Unless major changes are made, the burden on society will become impossible to bear.

And major changes *are* being made. For several decades after nursing homes began to increase in number and the population of frail elderly at home began to rapidly rise, not much progress was made in ameliorating the conditions faced by old people. But the situation began to change about 25 years ago, as more and more studies showed the factors that determine the disabilities of the aged. At the same time, the still small number of gerontologists were turning their attention toward ways to lessen the ravages of the added years that had been granted to the patients they were studying.

The field underwent a significant philosophical revamping in 1980, when James Fries, MD, a gerontologist at Stanford University, introduced a concept he called "compression of morbidity." By this, Dr. Fries meant the attempt to decrease the period during which any person is disabled. In general, most of us are now fated to endure a final period of many years during which we become ever more frail, with the trajectory of decline sloping downward more markedly after about 50. Dr. Fries hypothesized that measures could be taken to change the long, gradually drooping arc with a pattern that more resembles a slightly sloping horizontal line ending in a rapid drop-off shortly before death. If this was accomplished, he pointed out, "then lifetime disability could be compressed into a shorter average period and cumulative lifetime disability could be reduced." It is a concept very like the one Oliver Wendell Holmes wrote about in his poem "The Deacon's Masterpiece," in which he describes "the wonderful one-hoss shay [which] breaks down, but doesn't wear out."

You see, of course, if you're not a dunce
How it went to pieces all at once
All at once, and nothing first,—
Just as bubbles do when they burst.

There are those, of course, who would like to die "all at once," but others would prefer a short period of decline, so long as it is not at all like the agonized waning that so many suffer today. For large numbers of people in the developed world, the compression of morbidity is already beginning to happen. And there is evidence that it is within reach for far more of us.

Among the first steps in the process of change was the realization that physical frailty and not disease itself is the most important determinant of whether elderly people can care for themselves and remain vital contributing members of the community. The older the subject, the more important become such factors as muscle strength and bone density, particularly the former. Study after study has confirmed this observation, best stated by a team of gerontologists from the Netherlands in 1997, as follows: "[I]n the oldest old, loss of muscle strength is the limiting factor for an individual's chances of living an independent life until death." The operative words here are "until death." Imagine a world in which every very old person might continue to care for himself, enjoy his surroundings, and sustain his loved ones instead of the other way around. And all of this happy state of affairs would go on until near the time of his death.

As we move into early middle age, muscle strength begins to show its first evidence of lessening, largely because of our decreased ability to manufacture the necessary protein. The process of diminishment speeds up after 50, so that in the succeeding two decades we are likely to lose an additional 30% of our strength. Decreasing muscle mass means decreasing activity and its accompanying decreased stress on bones. Bone needs stressing forces; without them, there is no stimulus to remain thick and strong. To the amazement of many a physician with a large geriatric practice, it has proven to be not difficult to build the necessary muscle, even in the very elderly. It has consistently been shown that strength can be almost doubled within six or eight weeks in the oldest old, merely through a supervised regimen of high-intensity resistance training and weights.

But improving muscular power is only part of the answer—the brain needs plenty of attention too. It is by now no secret that continued intellectual stimulation is the key to avoiding many of the ravages of dementia and the apathy that steals the minds of so many institutionalized and homebound elderly.

Granted, all the reading and museum-exploring in the world are not likely to reduce the incidence of stroke, but such activities maintain synapses and probably encourage the development of new brain cells regardless of age. They keep us alert, mentally vital, and curious. If such factors as vigorous mental and physical exercise have such great benefits for the aged, how much better would the situations of old folks be had they started them earlier? Since the benefits of these actions, as well as others like quitting smoking, proper diet, and certain medical means (daily vitamins and perhaps a baby aspirin), have been known for years and have been adhered to by millions of Americans, their result should by now be statistically measurable. Have they, in fact, resulted in compression of morbidity? Evidence from eight separate surveys in which researchers followed older adults from 1982 until 1999 indicates that they have.[1] Not only that, but the proportion of people living in institutions for the debilitated elderly dropped from 6.8% of the population to 4.2% during the 17 years of the studies. There are plenty of journal publications demonstrating similar results, including a well-known one in which a large group of University of Pennsylvania alumni averaging 68 years of age at the beginning of the study postponed disability by an average of 7.75 years by exercising, avoiding cigarettes, and maintaining a normal body mass.

Though factors other than healthy lifestyles and improved attention to fitness must surely have contributed to these salutatory statistics, there can be little doubt that the relatively simple measures being recommended by gerontologists and their clinical colleagues, the geriatricians, play a major role in their attainment. There is reason for optimism in these figures, and there is hope that, in future decades, there will be fewer of the depressing scenes so frequently observed in our nation's nursing homes, where corridors are crowded with debilitated residents sitting aimlessly in wheelchairs and staring blankly at every passerby in uncomprehending bewilderment or stupor.

In a time of finite resources—and when have we lived in a time when resources were not finite?—we would do well to expend far more of our intellectual and fiscal capital on improving the quality of our later lives than on the self-absorbed and very likely fanciful goal of lengthening them beyond the 120 years allotted by natural selection. Biomedical science should aim to approach and even reach that figure, but not to go much past it. A larger population of the elderly would have much to offer society if they could remain sufficiently

1 Physical, cognitive, and sensory limitations in this group are beginning to decrease. Freedman, V. A., L. G. Martin, R. Schoeni (December 2002). Recent trends in disability and functioning among older adults in the United States: A systematic review. *Journal of the American Medical Association* 288(24):3137–46.

vital. Instead of spending vast amounts of money on today's version of Brown-Séquard's fantasy—rejuvenation clinics, where hormones and fetal cells are injected and fistfuls of antioxidants are swallowed—and instead of spending vast amounts of money supporting eager scientists who promise life everlasting by molecular sleight-of-hand, we should look at the aged around us and realize their need, and ours.

Some of those aged are already showing us the potential value of such investments in the quality of their lives. Despite the large numbers of debilitated people now in nursing homes, the older men and women we encounter as we go about our daily routines do look and act much younger than their predecessors did a generation ago. And we see far more instances of individuals whose lives are in many respects just as useful to society and rewarding to themselves as they were at the height of their working careers. They exemplify what can be done, if only the knowledge and the resources were made available to everyone.

Emanuel Papper, who died at the age of 87 in December 2002, was my closest and most admired friend. A U.S. Air Force medical officer decorated for heroism, he emerged from World War II with internal injuries sustained when his bomber was shot down, injuries from which he never fully recovered. Realizing that the field of anesthesiology was still in a relatively primitive state, he determined to do something about it, which he was able to accomplish by introducing the principles of basic science research into clinical care and the training of residents, an approach that he and only a few other pioneers brought to academic medical centers in the '50s. As chairman of the department of anesthesiology at Columbia University, he organized a program that attracted colleagues from all over the world; they came to learn his methods and bring them back to their own hospitals and medical schools.

After 25 years directing one of the United States' premier departments in his specialty, Manny Papper accepted the position of dean of the medical school at the University of Miami, where he remained for 14 years, retiring at the age of 70. Though he continued to teach residents and students in anesthesiology, his inquiring mind demanded new challenges, and he enrolled in the university's doctoral program in English literature, being awarded his degree when he was 76. The dissertation he wrote on the Romantic poets provided grist for his multifocused intellectual mill for the rest of his life, and he continued to pore over the works of early 19th-century authors and to acquire first editions of their works for his increasingly formidable collection. His intellectual interests were wide and included the study of religions, classical music, the masterpieces of painting of virtually every style, and a great curiosity about the undertakings of the wide variety of men and women with whom he came into contact.

Though essentially living a life of the mind, Manny Papper never neglected his body. He exercised regularly and strenuously, and was an avid tennis player until only a few years before his death. Anyone traveling to visit him in Florida was likely to be taken to the gym, and later invited to join him on one of his long, vigorous daily walks.

As Manny approached his mid-80s, he slowly developed shortness of breath and began to become more easily fatigued. A major abdominal operation to treat a life-threatening emergency caused by the old war wounds added to the magnitude of these problems, but he recovered rapidly and resumed his usual program of exercise, never having slackened off in his intellectual pursuits. When medical evaluation showed the breathing difficulties to be the result of a leaking heart valve, he resolved to undergo the hazardous surgery that would be required to repair it, because the respiratory limitations were infringing on his enjoyment of life, which still included a regimen of weight training and walking, though he had reluctantly abandoned tennis a few years earlier due to a shoulder injury.

The events of September 11, 2001, had reignited Manny's long-standing interest in Islam, which he began to study with the same intensity that he had always applied to his other pursuits, going so far as to organize a conference on the subject at the Aspen Institute, a global issues think tank, where he was a member of the board. Late one afternoon in early December 2002—a few weeks before the planned heart surgery—while exploring the Internet in the hope of answering some abstruse problem in Islamic history, Manny suddenly complained of blurriness of vision and then collapsed. He died in the hospital seven hours later, of a massive cerebral hemorrhage.

Of course, the story of Manny Papper represents an extreme example of what is possible, given the will and the wherewithal. He had the background, the personal drive, and the financial resources to live his years to the fullest. But the important lesson to be learned from his tale is that this kind of thing is possible, even if not very often to the extent that he was able to enjoy. What is required is the education of the public to understand the rather simple principles that prevent disability while increasing quality—and frequently the quantity—of life. And what is also needed is continued intensive research into the causes of disability and their prevention by the aging themselves, a category that each of us joins the moment we leave the womb. Better to spend our nation's money here than to waste it in the vainglorious search for immortality.

And if a few of our elder citizens believe that a more youthful outlook is attained by wearing a toupee or a wig, it is not for the rest of us to ridicule. Instead, we should consider that such stratagems are only a manifestation of the wish to remain young and are perhaps the first step toward buying the gym membership and the books that will really do the job. And as for the penile implant that my former classmate is planning . . . am I the only old codger who is beginning to think that it may not be such a bad idea after all?

AARON E. HIRSH

Signs of Life

FROM *THE AMERICAN SCHOLAR*

[Editor's Note: A biologist ponders the extent to which mathematics can describe evolution and ecological interactions.]

It was a four-word e-mail. "Booyah," it said. "Sincerely, Freight Train."

As soon as I read it, I stepped out of my office and hurried down the glass-walled corridor. Stealing a glance over Stanford's red-tiled roofs in the day's first light, I ducked into a room illuminated mainly by the cool fluorescence of a window-sized flatscreen. Ever since Hunter Fraser and I had begun working on an analysis of protein interaction and evolution, I'd been able to find him here at virtually any hour. He was now staring at the doorway, knowing his message would make me appear. Over the course of the night, he had inched his way down his seat, so that his head was propped against the backrest and his legs were stretched out before him, showing me the bottoms of his socks.

"Booyah?" I asked.

"Booyah," he replied, and nodded toward his screen. There, amid various spreadsheets of numbers and windows of programming code, were five pink rectangular columns, each taller than the last. To explain why this simple little picture merited "Booyah"—a word that had somehow worked its way into our collaboration to announce a result that was almost too good to be true—I need to back up a bit, and begin fairly near the beginning.

About a billion years ago, a population of single-celled animals, perhaps quite similar to the yeast that makes our bread rise, was divided into two groups. We don't know how they were split, but we do know that while they were separated, one of the groups accumulated so many genetic changes that when the two were reunited, they could no longer exchange genetic material. In other words, they had become separate species. One of these species had an evolutionary future full of big changes: it would discover that a single cell could band together with its offspring to make a multicellular organism, that separate cells could take on specialized functions, and that this could all be coordinated by the process of development. Ultimately, this species would evolve its way to wormhood, becoming a tiny soil nematode. The other species, meanwhile, would continue to work on being a yeast, outdoing its ambitious sibling in quantity if not in complexity.

Despite their obvious differences—one a worm that lives in the dirt, the other a yeast you can buy at your supermarket—these two species still have a surprising amount in common: they both still carry many of the genes that were present in that grandfather yeast a billion years ago. (Forgive me if we're rehashing high school bio, but we're almost there.) Each of these genes encodes a protein, a functional molecule that performs a specific job for the organism. And in many cases, the way this protein performs its job is by physically interacting with other proteins—actually touching them—to do things like pass a chemical message down a long molecular chain of command, or form large, multiprotein machines that execute complicated tasks. For the past few decades, molecular biologists have been identifying these interactions and reporting them to various databases, where you or I can read about them and play our own, maniacally complex game of connect-the-dots, drawing a link between each pair of proteins that interact, creating a reticulate web that represents most of the molecular contacts in the cell. In fact, it was Hunter Fraser's industriousness at this very task—finding and downloading massive amounts of data—that had earned him his nickname, "Freight Train."

In the weeks leading up to his e-mail, Freight Train and I had been working with a simple mathematical model of protein evolution. The model made a straightforward prediction: the more interactions a protein participates in, the more slowly it should evolve. Intuitively, the idea is that a protein with lots of jobs to do doesn't have much room for change: even the slightest alteration of its structure is likely to disturb at least one of its relationships with another protein. So if our little model reflected something real about the last billion years of evolution, we ought to see less change in multitasking proteins—the dots with lots of lines connecting them to other dots. But how could we possibly know how much a protein has changed over the past billion years? Well, if we took a protein in yeast and found a very similar protein in worm, we'd know that once upon a time, in that grandfather yeast, they were one and the same protein. So all we'd have to do to see how much evolutionary change had taken place is compare them. This is precisely what Freight Train had done to generate those pink columns: each one represented a set of proteins with a different number of interactions, and the column's height represented the amount of change that had taken place over the last billion years.

Using the chewed-up straw from his Pepsi as a pointer, Freight Train rested the gnarled white plastic on the tallest column.

"Zero to five interactions," he said. He added, with feigned pedantry, "Please note the surprisingly high rate of evolution."

His pointer moved to the next column. "Six to ten interactions. Note the significant reduction in evolutionary rate." And then he proceeded to the remaining columns, each shorter than the last: "Eleven to fifteen. Sixteen to twenty. And so on."

"It's unbelievable," I said.

"Nope," Hunter responded, as the straw went back between his teeth. "It's Science." Perhaps, but what seemed so improbable about his little picture was just how tidy it looked. How could millions of generations of worms wriggling in the dirt, and yeast replicating everywhere, through countless cataclysmic ice ages and scorching heat waves, through the drift of continents and the diversification of life—how could such chaos and change yield these five orderly columns? When Galileo suggested that "nature's great book is written in the language of mathematics," he was referring to physics and astronomy—elegant elliptical orbits, the reliable swing of a pendulum, the trajectory of light—not to worms in the mud, mutating and mating their way through a billion years of evolution. In this sprawling swamp of a science called biology, the short list of physical variables, such as force, mass, and energy, gives way to an endless catalogue of Latin taxonomy; prediction gives way to retrospective analysis; universal laws give way to idiosyncratic natural histories.

Yet when Freight Train plotted each protein's number of interactions against its rate of evolution, there emerged, through innumerable sources of noise and error, such a straightforward relationship: the more jobs a protein does, the more slowly it evolves, just as our little model had predicted. Granted, we were thinking statistically, not deterministically—we were modeling the average behavior of many proteins, not precisely predicting the rate of evolution of a single one. But could it be that if we think statistically and pick the right variables, we will eventually read even the evolution of a worm in the language of mathematics? Just how far will this rigorous symbolic system take us in our endeavor to understand an organism?

FIVE OF US are floating face down on the Sea of Cortez, and our shadows form a distorted rosette on the rocky bottom. At its center, in a small arena of sunlight, a vicious battle is taking place. It started a moment ago, when one of us shifted a rock and something pink, streamlined, and as long as a forearm spurted from beneath. A few meters away, the jet paused and gathered into the form of *Octopus veligero*. Fixing its penetrating stare on its assailant, it began shifting along the rocks, massaging its arms into crevices in search of a new retreat. It moved not like an animal, but like oil in water, its surface supple and undulating, and when it finally found an opening under a rock, it began to pour itself into the darkness. Its legs had slipped out of sight, but its pupils were still fixed on the five of us when its pink skin flushed suddenly red, then purple. A huge, pale tentacle shot from under the rock and wrapped across the eyes that had transfixed us. The purple body surged and pulled away from the seafloor. With it came the pale and swelling form of a second animal.

It is *Octopus bimaculatus,* two or three times the size of *veligero,* and apparently bent on destroying the small intruder. As we watch, the pale giant spreads itself into a terrible umbrella of tissue that envelops the small, purple creature. But when one of us dives to look more closely, *veligero* squeezes suddenly from its doom and retreats unsteadily, as though staggered to find itself alive. A furious red suffuses *bimaculatus,* a tentacle shoots out to apprehend the fugitive, and the battle rages on.

A few hours later, I stand beside a whiteboard in the seminar room of the Vermillion Sea Field Station. Behind the sunburned faces of a dozen college students, a window frame of weathered timber borders the almost abstract landscape of Bahía de Los Ángeles: a surface of sapphire interrupted only by the scorched desert islands that are scattered across the bay. Between two of these islands—the gently sloping Cabeza de Caballo and the small dome of Los Hermanitos—I can actually see this morning's dive site, where we watched the battle unfold and ultimately end with a severed tentacle, kaleidoscopic spasms of agony, and injured escape.

At first, the timing of that spectacle had struck me as opportune, for this was the day I was to lecture about interspecific competition—precisely the phenomenon we had watched play out so savagely on the ocean floor. And it was just this kind of intersection, I had thought, that had motivated our trip to Baja California in the first place. After all, when I wrote a description of this course, The Ecology, Evolution, and Natural History of the Gulf of California, it ended with this sentence: "Our hope is that daily experience of this magnificent marine ecosystem will lend life and specificity to the foundations of several biological disciplines, while these disciplines permit a deeper, more careful, and more rigorous examination of the ecosystem." Our students had just had the sort of "daily experience" we'd been hoping for.

But as I stand beside the equations I have written on the whiteboard and catch a glimpse of the morning's dive site, suddenly I'm not so sure. What I have written is a basic mathematical model of ecological interactions. It consists of a set of equations, one for each species involved, and it predicts the changes in the sizes of their populations over time. (A pair of such equations, one representing large fish, the other representing the little fish they eat, was first used by an Italian mathematical physicist named Vito Volterra to explain to his son-in-law, a marine biologist, why the cessation in fishing during World War I resulted in a striking increase in the abundance of big fish, but not small ones.) Despite their apparent simplicity, the equations actually predict a couple of counterintuitive effects, including the one addressed by Volterra. But as I try to convey the power of this simple model to the students—ordinarily an enthusiastic and engaged group—they seem unmoved. A few of them glance out the window. Then I, too, catch sight of the sea, and a moment later I suggest we take a break.

As the students grasp the announcement of their liberation and move uncertainly toward the door—like prisoners told to leave, it seems—one of them turns to offer me a charitable expression of interest. When I lament my failure to explain why these equations were so incisive, he says something that at first sounds utterly ridiculous: "Maybe we would have gotten it if you'd written one equation in red and the other one in purple."

After a moment, I realize that, in a sense, he's right. The mathematical language that once seemed incomparably powerful—capable of distilling the mess of evolution to a few key variables and precise, general rules—now seems pathetically deficient, inadequate to the task of capturing even the thinnest slice of what matters in nature. Today, the red of fury and the violet of dread seem far more useful symbols for understanding life in this bay.

IF YOU HAVE EVER heard of Bahía de Los Ángeles, it was probably from John Steinbeck. He and Ed Ricketts, a zoologist, hard drinker, and womanizer from Cannery Row in Monterey, passed through here on an eight-week chartered voyage that was variously collecting trip, inebriated frat party, and errant

romantic quest. In a world without dragons, one can always collect nudibranchs, and the expedition's trials and euphorias are vividly recorded in *The
Log from the Sea of Cortez,* Steinbeck's paean to the virtues of male bonding,
fixing engines, and finding strange creatures in the sea. Insofar as they were
doing science, Steinbeck and Ricketts were doing descriptive invertebrate taxonomy and biogeography—cataloguing the wild variety of intertidal life and
where it was found. This kind of natural history is very different from the sort
of science I was doing with Freight Train or the ecological modeling that fell so
flat in my lecture: one is an effort to describe the diversity and distribution of
species; the other is a search for simple mechanisms and laws that underlie the
complexities of organic life. While one revels in the riot of particulars, the
other seeks precise universals. Despite the differences, however, Steinbeck was
also concerned with the relationship between Doc Ricketts's scientific descrip-
tions of nature and his own experience in the Sea of Cortez.

> For example: the Mexican sierra has "XVII-15-IX" spines on the dorsal fin.
> These can easily be counted. But if the sierra strikes hard on the line so that
> our hands are burned, if the fish sounds and nearly escapes and finally comes
> in over the rail, his colors pulsing and his tail beating the air, a whole new
> relational externality has come into being—an entity which is more than the
> sum of the fish plus the fisherman. The only way to count the spines of the
> sierra unaffected by this second relational reality is to sit in a laboratory, open
> an evil-smelling jar, remove a stiff, colorless fish from formalin solution,
> count the spines, and write the truth "D. XVII-15-IX." There you have
> recorded a reality which cannot be assailed—probably the least important
> reality concerning the fish or yourself.

To be sure, Steinbeck's spine count is no exemplar of concise mathematical
analysis. "D. XVII-15-IX" may not seem at first to tell you much about the
sierra's spirited fight, but in the context of evolutionary biomechanics, it tells
you something about where this fish came from and how it fits into a story in
which the tail became deeply forked, providing tremendous thrust; the upper
jaw was fused to the skull to support a powerful bite; and the pump for pushing
water over the gills was replaced by incessant, rapid swimming—all of which
do, after all, tell you something about the sierra's spirited fight. But even if
Steinbeck is an imperfect representative of rigorous science, the sentiment he
voices is nonetheless exactly the one that took hold of our lecture room that
day in Bahía de Los Ángeles. Must a scientific description of life be so lifeless?
And not only dead, but also trivial? How to reconcile one day's sense that a

mathematical model could represent a deep and universal rule with another day's realization that a model told us far too little about the creatures we watched just hours before?

A SIMPLE MECHANISM WAS discovered beneath the diverse particulars of natural history when Charles Darwin scrawled "I think—" in his notebook and then sketched a small diagram of the operation of natural selection. This concise model had not come fortuitously—like the mythic apple knocking Newton's head—but had been steadfastly pursued by a naturalist who strived to place the soft and messy science of life on the same firm and lofty footing as physics. Indeed, Darwin's desire to confer Newtonian status on his new laws of evolution is evident in the final sentence of *The Origin of Species:*

> There is grandeur in this view of life, with its several powers, having been originally breathed by the Creator into a few forms or into one; and that, whilst this planet has gone cycling on according to the fixed law of gravity, from so simple a beginning endless forms most beautiful and most wonderful have been, and are being evolved.

Gravity? What's the law of gravity doing there, except perhaps reminding us that nature is rigorously lawful? Just as the planets move according to Newton's laws, Darwin seems to suggest, so life is transformed according to his own. Yet Darwin also seems to recognize that a scientific understanding of evolution threatens to make organisms seem lifeless or trivial. Why else would he close *The Origin* by asserting—more than a little defensively—that there is grandeur in this view of life? He places life's beauty and wonder, even a breath from the Creator, alongside the law-abiding and precise motion of planets through space, and with that concludes the work that bequeaths to us the very juxtaposition we all felt that day in Bahía de Los Ángeles. But if this work is indeed the origin of the tension between a precise, law-based depiction of life on the one hand, and a naturalist's experience of vivid particulars on the other—and if Darwin himself is acutely aware of the conflict—then we must wonder: Does he really believe that there is grandeur in his view of life? Or is he merely trying to soften the metaphysical blow of supplanting God and wonder with scientific law? Does he too find that a scientific depiction of life renders it as lifeless as a formalin-soaked fish? Or is his sense of beauty, wonder, and a breath from the Creator somehow compatible with his scientific depiction of nature?

In some ways, it is strange that Charles Darwin should have longed for concise mechanisms and universal laws to underlie the science of life. He adored collecting, and expertly practiced the same kind of descriptive natural history that Steinbeck and Ricketts would pursue—albeit somewhat drunkenly—a hundred years later. The author of *The Origin,* we shouldn't forget, was also the man who wrote four monographs on barnacles, as well as dozens of other texts, from *Geological Observations on Coral Reefs* to *The Formation of Vegetable Mould, through the Action of Worms.* And while his work on barnacles is often portrayed as an intellectual refuge from his more momentous discovery, it's hard to believe this is the whole story, since he'd been investigating the minutiae of organisms, mastering their diversity and memorizing their peculiarities, ever since he was an adolescent. In short, Darwin so adored the prolixity of Nature's Book that it is surprising he should have struggled so doggedly to describe it in terms of a simple set of laws.

It seems even stranger that Darwin should have aspired to found a nearly Newtonian science when we recall his confession that the rigorous study of rule-based systems—namely mathematics—was not his strong suit. But if he was weak in math, he was a master of the written word, and he wielded it with no small flourish. Take this passage from *The Origin,* where he offers a metaphor for the process we would call (rather austerely) macroevolution:

> As buds give rise by growth to fresh buds, and these, if vigorous, branch out and overtop on all sides many a feebler branch, so by generation I believe it has been with the great Tree of Life, which fills with its dead and broken branches the crust of the earth, and covers the surface with its ever-branching and beautiful ramifications.

How is it that this man, so fond of collecting, so encyclopedic in his knowledge of natural history, so uncomfortable with the symbolic systems of math but so powerful with words—how is it that he was the one to introduce concise mechanisms and universal laws to the study of life?

AN ANSWER, PERHAPS, is that despite Darwin's aspirations, natural selection is in fact very different from Newton's laws. It is indeed law-like in the sense that natural selection does, without exception, underlie the evolution of living beings. But it is a law with lawless outcomes—a universal set of rules that play out differently in every species and every context. They do not yield a billion years of elliptical orbit, but rather start with a single, simple organism, and in a billion years fashion yeast or worms, *Octopus bimaculatus* or *Homo sapi-*

ens. So while natural selection does impel the growth of every bud on Darwin's vast and ramifying tree, there is no telling where new branches will sprout, which direction they will grow, or how they will flower.

Because of this property—their sheer creativity—Darwin's "laws" do not encapsulate the transformations of life in quite the same way that Newton's laws capture the motions of objects. They render evolution intelligible, but not predictable or reducible. That is, to understand the evolution of any single organism, the sparse outline provided by Darwin's model must be filled in with long passages of particularities—details of the organism, its development, its mutations, its ecology, the events that befell its ancestors. Only then does the simple, universal model cast light on any particular organism. And only when the laws serve to elucidate the organism does a study of the organism reveal the action and significance of the laws. Perhaps this is why it was someone like Darwin, whose mind veritably teemed with the details and histories of organisms, who detected the mechanism underlying the evolution of life.

If the simple mechanism that Darwin discovered were not so wildly creative—if his rules could reduce the organism, predict its form and behavior, successfully encapsulate it—then the elegant, scientific language of key variables and the precise relationships among them could hardly fail us in our contemplation of nature. But of course this language does fail us. That is precisely what happened that day in Bahía de Los Ángeles. For the purpose of sharing with my students the spectacle of animals engaged in battle, their anger and terror, their versicolor skin and the writhings of their molten forms—for this purpose, my calculus was of little use. The Book of Nature may well be written in the language of mathematics, but it is written in many other languages as well, and passages inscrutable in one may be legible in another.

Does this mean that passages deciphered in concise mathematical terms are necessarily a dead read, drained of life by the very act of cutting to the mechanisms and laws beneath? If it is the sheer creativity of natural selection that gives rise to the tension between a scientific reading and our experience of the organism, perhaps one way to dissolve this tension is to keep that very creativity in mind. That may sound at first like some kind of koan, but what I mean is only that when we consider the relationship between Darwin's simple mechanism and its wildly varied outcomes—when we remember that the relationship is not one of reduction or encapsulation, but rather one of intelligible creation—the tension between the observation of the organism and the contemplation of simple scientific models seems more or less to dissolve.

I once spent fourteen days looking for a leopard, but in the end I saw only a pugmark left in the mud. It was nothing but five small puddles, but it was enough to freeze me—to halt me with the sense that something powerful and elusive had walked the trail ahead of me. The picture on Freight Train's monitor—the basic rule that proteins with more interactions evolve more slowly—was but a mathematically legible footprint left by an extraordinarily creative mechanism. And just as an indentation in moist dirt can be charged with the presence of an animal, nearby but invisible, a mathematical footprint can be infused with a sense of the wild, varied, billion-year procession that came this way before you. The traces on the trail are countless, of course—from the habits of worms in the dirt to the patterns we observe in their proteins, from the spines of a sierra to the eye of an octopus—and the more of them we notice, pause beside, and inspect, the sharper grows our sense of the laws that made them all. Perhaps, then, these laws need not deaden our view of life; perhaps our view of life can vivify these laws.

IAN PARKER

Reading Minds

FROM *THE NEW YORKER*

[Editor's Note: Attempts to communicate with people "locked in" by total paralysis redefine the concept of putting thought into action.]

One Friday morning in July, 2002, Niels Birbaumer, a neuroscientist from the University of Tübingen, in southern Germany, was driven from his hotel in the center of Lima, Peru, to a house protected by a half-dozen armed guards in a gated suburban enclave. He was accompanied by two colleagues—Herta Flor, a professor of neuropsychology, who is his girlfriend, and Thilo Hinterberger, a physicist and computer programmer—and by José Luis Palomino, who runs a company in Lima that distributes communication devices for disabled people. With the assurance of a regular visitor to the house, Palomino led the group through a split-level main room, filled with zebra-skin cushions and potted orchids, to a den decorated with sporting trophies, a stuffed deer's head, and a two-foot bottle of Teacher's whiskey. Sliding smoked-glass doors opened onto an L-shaped swimming pool in a bright, spongy lawn. A game of tennis could be heard from the other side of a tall hedge.

Palomino had set up a makeshift laboratory in one corner of the room. Two desks were placed at right angles: a large television monitor was set up on one, and on the other a computer was connected to a small amplifier. The three visitors from Germany plugged in their own laptop computer, and ran tests on the equipment, while politely accepting snacks brought by shy, silent maids. Esteban Ripamonti Aguad, nicknamed Polo, a jumpy, slightly unkempt man of twenty-five, who is the nephew of the owner of the house, came into the room. "We're very happy to have you here," he told Birbaumer, in Spanish. "We've been counting the days." He spoke quickly, underlining his words with whistles and darting hand movements.

Sometime after midday, there was a whisper—"Mr. Elías is coming!"—and through the glass doors one could see a wheelchair being guided very slowly down a ramp by three nurses and a doctor. Polo's uncle, Elías Musiris Chahín—the wealthy owner of a casino and two fabric factories—sat motionless in the chair. He had a scarf over his mouth, another over his head, and blankets covering his body. He was wearing wraparound dark glasses. On his feet were oversized slippers designed to look like leopard's paws, finished with cloth claws.

Once Musiris's procession reached the room, a nurse took off his glasses and scarves, and the extent of his paralysis became apparent. He was breathing with a respirator. His left eye was open only because the eyelid was held in place artificially, and the eyeball was stationary. His mouth, which he cannot move, was open very wide, as if at the midpoint of a yawn. Now and then, a nurse delicately used a damp cotton pad to push her patient's lower eyelid upward, to approximate a blink, and cleared saliva from his mouth with a small suction tube.

To use the blunt medical term, Musiris is "locked in": he is unable to communicate with the world, although it is assumed that his senses and his intellect are intact. In 1996, when he was fifty-three years old, Musiris received a diagnosis of a form of amyotrophic lateral sclerosis (ALS, or Lou Gehrig's disease), which is incurable. In the winter of 2002, he completely lost deliberate movement in his eyes, and, with it, the ability to indicate "yes" and "no." His thoughts became unknowable. Nobody fully locked in has ever answered a question.

Musiris's wife, Estrella, joined her husband. She is a tall, broad-shouldered woman in her mid-fifties, with a face dominated by dark-red lips. She made sweeping gestures of welcome, and gave everyone kisses, but she appeared weary in the role of expansive glamour. As Birbaumer later learned, she has been at odds with Musiris's two adult sons from his first marriage, largely over management of his businesses. It was also reported in the Peruvian press that she had argued with her sister-in-law over the origins of Musiris's condition— Estrella charged the sister-in-law with witchcraft, and was in turn accused of poisoning her husband with chemicals used to cultivate orchids. Meanwhile,

Ripamonti, who is her nephew, has embraced the role of dutiful son. A few years ago, he moved in and took charge of Musiris's care. Depending on whom you ask, he either filled a vacuum that the dispute with Musiris's sons had created or executed a coup d'état. In 2002, he started saying that he could communicate with his uncle by reading how his pulse rate changed. He took questions from Musiris's family and business associates, and returned with answers after some hours in private with his uncle. (In the opinion of Birbaumer, who was never allowed to witness the process, the technique was "impossible"—Musiris was on artificial respiration, so he would not have sufficient control of his heart rate—and Ripamonti was guilty of wishful thinking, at best.)

Birbaumer knows that good science can happen in makeshift settings—it can flex in response to place or the impact of a new experimental thought—but for a moment he looked bemused by his surroundings, and by the size of the crowd now growing in the poolside room. He went over to Musiris, and, in a gentle alternative to a handshake, touched the back of his pale hand with three fingers, and spoke to him quietly in English and in broken Spanish. Musiris was then placed at a point about six feet in front of the TV screen. Birbaumer measured his head, and marked positions on his scalp with a red felt-tip pen. He glued electrodes to the spots, and connected them to the amplifier. "Don't try too hard," he said encouragingly to Musiris, who had been introduced to the equipment in the weeks before Birbaumer's arrival. He added, *"Vamos."*

BIRBAUMER HAD FLOWN to Lima to attempt the first-ever conversation with a man fully locked into his body. He is a leading figure in a new field of research seeking unconventional lines of command between human brains and the world beyond them. Brain-Computer Interface (BCI) systems, as they are sometimes known, make an unprecedented proposal: to separate human communication from human muscle, to give thought the power of action. Normally, a brain must call on its body to make something—anything—occur. But BCI imagines a short circuit: a brain wants something to happen, so it happens, without any need for a mouth or a mouse or a remote control.

BCI research has taken two paths: one proposes to implant electrodes in the brain in order to pick up electrical activity there (with one exception—research that began at Emory University, in Atlanta—this has only reached the stage of animal experimentation); the other uses electrodes at the scalp, reading a fuzzier and therefore less useful signal, but without the risks of surgery. The field carries enough promise of dazzling future applications to have created an international network of bad feeling between—and within—these two groups. Birbaumer works in the noninvasive school, as does, most notably, the neuroscientist Jonathan Walpow, who heads the BCI project at the Wadsworth Center of the New York State Department of Health, in Albany. No one doubts the seriousness of Birbaumer's work—he publishes in prestigious journals—but, as he happily acknowledges, its importance lies less at the level of pure scientific breakthrough than in its ability to bring assistance to actual patients. So it's with a little swagger—a defensive arrogance—that Birbaumer calls the equipment that was now hooked up to Musiris's head a Thought Translation Device, or TTD.

The nurse who was attending to Musiris's eye stayed at his side, and the rest of the medical team took seats on wicker furniture. Birbaumer sat at the desk. On the monitor in front of Musiris, a white dot began to move across the screen, as if it were tracing a record of seismic activity. Its progress from left to right was steady—an eight-second journey, repeated as soon as it was finished—but it leapt up and down on its vertical axis. Musiris was being shown a part of his electroencephalograph, or EEG, reading, filtered and amplified. The up-and-down movement was a live reflection of his brain activity: a reading of the incessant, faint electrical buzz of a hundred trillion synapses; a pattern that differs if one is awake or asleep, calm or panicked. Each time the dot made its left-right journey, one half of the screen was pink—sometimes the top half, sometimes the bottom, in a random sequence. Musiris's task was to change the electricity of his brain by changing the thoughts inside it, and to use this to control the ball, to keep it in the half of the screen that was pink—the required direction being reinforced by a recorded voice saying *"arriba"* or *"abajo,"* up or down. He was being asked to play a crude video game using only his brain.

"To take the light up, maybe try to imagine you're about to fire an arrow—a preparation for action," Birbaumer suggested. "To bring it down, think of something in the future. A projection." Everyone watched the screen. The room was silent except for the whirring noises of Musiris's respiration and waste-removal systems. *"Arriba,"* the disembodied voice said, and a light began bouncing its way across the screen. Musiris's audience willed it upward. It stayed up, and Musiris scored a "hit"—that is, the dot spent more than half of its left-to-right journey in the correct half of the screen. A soothing arpeggio chord sounded. There was another *"arriba,"* but this time the point of light dipped too much, and there was no rewarding chord. Then *"abajo,"* and the light went down—another hit, and so it went on, with almost as many misses as hits. Birbaumer shook his head. A random response would generate a fifty-percent score; Musiris was barely managing even this, so no one could confidently regard the hits as deliberate. But the power of placebo is strong, and when the light did obey the instructions—*"abajo"* or *"arriba"*—it was hard not to read this as an act of furious determination, and to follow the light's unsteady movements as one would a dumbbell above the head of a struggling weight lifter.

The experiment ended within an hour. The medical team wheeled Musiris out of the room. Birbaumer, who, from experience, believes in the importance of reaching a patient before he or she is fully locked in, did not look optimistic. "What you haven't learned when you're young, you haven't learned," he said, quietly. "If you haven't learned this before you're locked in—" He hesitated before adding, "Maybe we've come too late."

Birbaumer and his colleagues were summoned to lunch. Ceviche and squid had been laid out on a long glass table. Musiris sat at one end of the table, his nurses standing behind him, his open mouth covered by the scarf. (He has not been able to eat since 1998; he is fed liquid nutrients, by tube, into his stomach.) At the other end of the table, directly in his line of sight, a television set showed live security-camera images beamed by microwave from his casino, a few miles away. Estrella Musiris pressed food and wine on her dozen or so guests, who—besides Polo, Palomino, and the scientists—now included Eddie Thornberry, who is Musiris's business manager, and two of Estrella's sisters. One of them, Polo's mother, was wearing a jaunty sailor's hat; the other passed around a set of Polaroid photographs that showed her making big wins on slot machines. (They proved that gambling was financially rewarding, she said.)

Midway through lunch, a servant carried in Estrella's new pet monkey, Lorenzo, who was wearing a red cardigan. Lorenzo was introduced to Birbaumer, then made a shrieking dive for the food, and was taken off to the kitchen. I asked a family friend at the table what he thought Musiris's first words would be if he regained the ability to communicate. He smiled, and said, "Kill the monkey."

WHEN PEOPLE ASK Niels Birbaumer about the mental life of a person locked into his or her body, he often mentions a favorite book, Dalton Trumbo's 1939 novel, *Johnny Got His Gun*, which is narrated by an American soldier who wakes up in a hospital bed to discover that his legs, arms, and face have been blown off. The soldier is deaf, dumb, and blind. "He was a dead man with a mind that could still think," Trumbo writes. After several years, the soldier comes to realize that by nodding his head he can tap out messages in Morse code. When military commanders are brought to his bed to hear what he has to say, he asks to become a travelling exhibit, demonstrating the ugliness of war. They reject the idea. They sedate him and leave him. "He was the perfect picture of the future and they were afraid to let anyone see what the future was like."

Birbaumer finds in the locked-in syndrome a kind of revolutionary model, with patients oppressed—as he sees it—by neurologists who encourage them to suppose that life on artificial respiration is unlivable; and, on occasion, by family members who discover the advantages of a silent and compliant relative ("the perfect spouse!" he says). He is a liberator, giving speech to the silenced. But Birbaumer is also sometimes mesmerized by the idea of a life of contemplation. "There are days when I think, Oh, I could not stand being locked in," he told me during the summer of 2002, shortly before he flew to Peru. "And then other days when, well, I imagine just being able to think, do nothing but think."

We were in Tübingen, where he runs the university's department of medical psychology. The window of his third-floor office looks past steeply gabled houses toward the River Neckar, where tourists were being punted upstream and down in long, flat boats. Birbaumer, who prefers more dynamic surroundings, and whose own volubility contrasts with the silence of his patients, sat slouched in an armchair, and frequently answered a tiny cell phone in his breast pocket, speaking in German, Italian, and English. There was some sighing and shrugging. Birbaumer, a slight, dark, well-dressed man of fifty-seven, his hair turning gray, devotes much of his day to exasperation at one thing or another: his allegedly bland, overambitious PhD students; the "bullshit" research being done in his field; the "idiots" in Europe and the United States who deny him funding. He is brusque but good company.

In Germany, Birbaumer is a well-known public intellectual who contributes—in an amused, reckless, and sometimes cutting way—to television debates on social and medical matters. His political stance points to a background in the student upheavals of 1968; born in Czechoslovakia and brought up in Austria, Birbaumer is a graduate in psychology from the University of Vienna, and was a professor there until he was fired for agitating against the teaching of an allegedly reactionary senior colleague. He moved into Germany's wrecked postwar scientific community, and was promoted quickly (some think too quickly), becoming a full professor of psychology at Tübingen at the age of twenty-nine. He has been married to the same woman twice, finally separating from her in 1991, and he has an adult son from another relationship. Twelve years ago, he met Herta Flor, who is a leading researcher in the study of pain at the University of Heidelberg. (Her recent experiments showed that a patient's sense of pain can increase threefold when his or her spouse is in the room.) Today, he spends weekdays in Tübingen, then, each weekend, he and Flor drive to a house in northern Italy, where they grow olives and keep pigs. He is a member of what used to be called the Italian Communist Party.

Birbaumer's PhD thesis described how the brains of blind people make structural compensation for the loss of sight; he later studied the mental activity of musicians at the moment before they play a note. From early in his career, he has worked with the brain waves known as slow cortical potentials (SCPs). In the range of signals that make up the human EEG, SCPs are unusual for occurring over a period of seconds, rather than milliseconds, and they tend to accompany all kinds of human activity; they are not linked to particular motions or emotions. An SCP reading is a fairly reliable indication of general brain business. In work that led to Birbaumer's winning a Leibniz Prize—Germany's best-endowed research award—in 1995, he trained epileptics to fend off an impending seizure by adjusting their SCPs. Birbaumer might have called this process "biofeedback" had the word not been tainted in the nineteen-sixties and seventies by an academic scandal involving the falsification of data, and by "white-robed meditation crap," as he puts it. ("I now call it 'physiological regulation,'" Birbaumer told me. "With this, you impress the medical profession, and the psychologists don't know what you mean. So you don't get into any trouble.") Using feedback carried from electrodes on the scalp, shown in a visual display—a ball rising and falling on a computer monitor—fifty epileptic subjects, all of whom were unresponsive to drug treatments, learned over the course of thirty-five sessions how to send their brains into a state of SCP "positivity." When a seizure loomed, they would alter their SCPs. They could think their way out of danger. Birbaumer found that, on average, over the following months, his patients were able to reduce the frequency of their seizures by half; about a third stopped having seizures altogether.

In the late nineteen-eighties and early nineties, the field of BCI was beginning to emerge in Europe and the United States—driven in part by cheap, powerful computers, and by a louder lobby for disabled people. (Much of the neuroscientific knowledge had been in place for a while: electrical brain signals were discovered in 1875, and the human EEG was first recorded in 1929.) It was clear to Birbaumer—and others—that if one could change one's SCP reading at will, as his epileptics had done, and if this change could be picked up at the scalp efficiently, and amplified, then this technology might serve as a communication device for people with severe disabilities. The money awarded with the Leibniz Prize—$1.5 million—gave Birbaumer an opportunity to start a new experiment. In 1995, he began training paralyzed people to write with their brains.

THE LOCKED-IN STATE can be mistaken for a coma, and a coma can be mistaken for a locked-in state. A locked-in diagnosis (the term was first used in 1966 by the American neurologists Jerome Posner and Fred Plum) means that although a person has been abandoned by his or her body, the brain still functions. The most clear evidence for this is usually some residual voluntary eye movement, a deliberate response to "Blink if you hear me." A comatose patient does not blink; in tests, locked-in patients have responded with eye movements that show they have suffered no significant loss of language, memory, or intellectual function. In two rare cases of deliberate total self-paralysis—unconnected experiments in the mid-nineteen-forties to test the possible usefulness of muscle relaxants in surgery—scientists in London and Salt Lake City found that when briefly locked in they could see clearly (if someone pulled up one of their eyelids), hear, feel the pain of pinpricks or adhesive strips pulled from their skin, and experience terror ("I felt I was drowning in my own saliva," one later reported).

A French magazine editor, Jean-Dominique Bauby, was almost fully locked in after a stroke in 1995, and used his left eyelid to dictate a wry memoir, *The Diving Bell and the Butterfly*, published just before his death, in 1997. "I have known gentler awakenings," he wrote. "When I came to that late-January morning, the hospital ophthalmologist was leaning over me and sewing my right eyelid shut with a needle and thread." The book explains the method of its composition. An assistant at Bauby's hospital bedside would run through the alphabet—rearranged according to the frequency of each letter's use in French—and Bauby would blink at the letter he needed, and so describe swooping mental journeys he took to escape from his body:

> I reshoot the close-ups for *Touch of Evil.* Down at the beach, I rework the dolly shots for *Stagecoach,* and offshore I re-create the storm rocking the smugglers of *Moonfleet.* . . . I am the hero of Godard's *Pierrot le Fou,* my face smeared blue, a garland of dynamite sticks encircling my head.

According to the medical literature (which includes a description of an American man who, at the time of his death, in 1990, had been locked in for twenty-seven years), the syndrome has at times been induced by a brain tumor, by alcohol abuse, and by a visit to a chiropractor after a car accident. But it is most commonly associated with brain-stem strokes (as in Bauby's case), extreme spinal-cord injuries, and neurological diseases like ALS. As a result of recent medical progress—in particular, refinements in methods of artificial respiration—the locked-in population is now growing. There are no statistics, but Birbaumer guesses that there are twice as many locked in patients in the world as there were ten years ago: he estimates that there are perhaps ten thousand in Germany and twenty-five thousand in the United States. A small proportion of them, perhaps one or two per cent, have survived beyond any lingering eye movement (and beyond any slight sphincter movement, which is often the last control to be lost), to reach a place of complete separation from the world. Among them is Elías Musiris.

THE SON OF PALESTINIAN immigrants, Musiris made his fortune in acrylic and cotton cloth. "He was a successful man, handsome, very sexy," Estrella Musiris told me. "You remember the actor Omar Sharif? This man, he was Omar Sharif." He married his first wife in the late nineteen-sixties. They had two sons, Arturo and Javier, and eventually divorced. He met Estrella in the eighties, but, because his divorce was not final, they did not marry until just before he became ill. "It was love at first sight," she says. "True love, to the end." She and Musiris would sometimes fly to Las Vegas for the weekend. He loved gambling and horse racing. He owned a stable of horses he called Jet Set, and was elected the director of the Jockey Club of Peru. On the walls of a back room in his house, a group of framed photographs show Musiris at the track over two decades, wearing binoculars around his neck, sometimes flanked by friends in dark glasses and dark shirts, posing with a cup or with a winning horse. Over the years, a little paunch appears under his well-cut jackets, a thin mustache comes and goes. His sons grow up from neat boys in shirts and ties to handsome, privileged teenagers in flared trousers. In later photographs, Estrella is at his side, always tilting her head in a wistful way.

In 1996, when Musiris's ALS was diagnosed, he was optimistic that a treatment might be developed in time to help him; he visited at least ten specialists in the United States. Many ALS patients experience a long period of slowly worsening symptoms, but the disease ambushed Musiris. Within a year, he could no longer walk or talk, and he was already on artificial respiration when, in 1997, he was introduced to José Luis Palomino. Then in his mid-thirties, Palomino was a former Navy intelligence officer who had once been active in the government's bloody struggle against the Shining Path guerrilla movement. After a period of self-imposed exile, he returned to Peru, working first as a security consultant and then as a computer-systems adviser specializing in tools for paralyzed people. When they met, Musiris was communicating with an infrared device fixed to his head: by moving his neck, he could direct a point of light on a grid of letters in front of him. "He was losing movement," Palomino told me, remembering his first meetings with Musiris. "I was the only person who told Mr. Elías that there was a strong possibility he would be locked in." Musiris brushed his concerns aside. "He'd say, 'Have another coffee, have a dessert.' "

Palomino set up a system incorporating a beam of infrared light pointed at Musiris's chin, and EZ Keys, a word-prediction program that facilitates typing with a single movement. Then his chin movement became unreliable, and Palomino linked EZ Keys to Musiris's static charge, which he could adjust by touching his teeth together. This control was also soon lost. By the end of 2000, Musiris was able to communicate only by glancing to the left or the right. Palomino says that, using this method, "he told me, 'Keep up the communication, whatever it costs. Buy everything. I will try everything.' "

By the beginning of 2001, it was hard to make sense of Musiris's eye movements. By the end of the year, he was fully locked in. After reading about Birbaumer on the Internet, Palomino contacted him by e-mail, and last April he and Ripamonti flew to Tübingen for a meeting. Birbaumer is emphatically not in the business of marketing his TTD—he has a leftist's disregard for commerce—but he was struck by Palomino's competence, and was willing to give him a copy of his software, help him buy the hardware, and show him how he was using the combination with patients.

Palomino began training with Musiris, and Birbaumer studied his initial results. They were not spectacular, but because there was some sign of response (and because Birbaumer's brother lives in Paraguay) he agreed to spend three weeks with Musiris in Lima, accepting expenses but no fee. "It's a chance to communicate with a real locked-in patient," he said. "Maybe."

WHEN I FIRST MET Niels Birbaumer, in 1999, he was four years into the BCI experiment, and seemed confident: he had made progress with several patients at different stages of paralysis, and had recently published an article in the journal *Nature* about his success with Hans-Peter Salzmann, a former lawyer in his forties who had been disabled by ALS. Salzmann lives in Stuttgart, about half an hour north of Tübingen. Birbaumer met him in 1996, when he was already paralyzed to the point where he controlled only a single muscle under his left eye; with this, he had a Bauby-like means of communication. Birbaumer had found his patient to be smart, methodical, and a little formal—

"Prussian," as he puts it. (Even today, Salzmann, who can still communicate with his face muscle, has never used the more familiar form of "you"—*du*—with Birbaumer.)

Some people find it easier than others to influence their EEG; Birbaumer has discovered no correlation with intelligence or age (although depressives seem to do better than average). Barbara Wilhelm, a research student in Birbaumer's department, told me that she regularly scores a hundred per cent with her SCPs. "To make the ball go up, I think of the 'glory of love,' " she said, with a faint blush. "I just think of that phrase. And, to make it go down, I think of peace and calm, of peace and freedom—the end of the Arab-Israeli conflict." Salzmann, a former athlete, made the ball rise by imagining himself crouched on the starting line of a race. Salzmann did not find it easy, but, over time, he improved his technique; he was persistent and highly motivated, and after a year or so he had begun to reach scores of seventy per cent. At this level, he could advance to the next, crucial, stage: a spelling program written in Birbaumer's department. When combined with a monitor, electrodes, and an amplifier, this constituted the TTD. Salzmann saw a screen with half of the alphabet above the horizontal line. Did he want one of these letters? Yes or no? If yes, he had to send the ball up. If he sent the ball down, the other half of the alphabet would appear. Whichever letters were chosen, they would split again into two, and Salzmann would choose again, and so on until he had reduced the alphabet to a single letter, a process that usually took several minutes. (There was a third screen at each stage, where he could choose to move back to correct an error.)

For the first months, Salzmann was asked to reproduce letters, or strings of letters, given to him by his trainers. Then came the point when he was invited to spell on his own. Andrea Kübler, a research student then working with Salzmann, once described to me a landmark day in 1997. Salzmann produced the letter "A," then "B," "and then 'R.' And I thought he'd been trying to write *aber*—'but'—or *abend*—'evening'—and he'd written the wrong letter, and hadn't managed to delete it." But then he produced "A" and "K." Relatively speaking, the word spilled out of him: "ABRAKADABRA." Soon afterward, Salzmann wrote a letter to Birbaumer thanking him for his work and inviting him to a party. The letter was printed in *Nature*.

I visited Salzmann a few years ago, and again in July of 2002. He lives in a ground-floor apartment in a modern building, and he is assisted at all hours by nurses and by young people who have chosen to do community service rather than enter the military. In a sunny office next to his living room, he was sitting at an angle of about forty-five degrees in a wheelchair, with his head propped up straight; his hands were in his lap. Tubes supplied him with air and food. There was a novel by Elias Canetti on a stand and on the wall a black-and-white photograph of Salzmann, taken years before, at a picnic in a field of high grass. He gave me a demonstration of a new program developed by his Tübingen researchers: he could now surf the Internet (or at least a pre-selected part of it) using his brain. I watched him visit the site for C. H. Beck, a legal publishing house, and then choose, from the home page of the newspaper *Süddeutsche Zeitung,* a story about the difficulties facing Rudolf Scharping, who was then the German defense minister.

Salzmann is still able to make himself understood without BCI, if an experienced interpreter looks directly at him from a few inches away, offering him letters. Through a Tübingen research student, Ahmed Karim, Salzmann told me that he was glad of the warm weather and of the chance he had that morning to sit outside with his nurses. He told me that his habits of mind had not changed with his illness, but he added, "I have a more vivid imagination when I write, and I think that in general I've become more attentive. I concentrate and I try to memorize more." He said that Bauby's description of the locked-in state corresponded to his own experience, although he dislikes the phrase. "As long as I can express myself, I'm not locked in," he said.

A FEW DAYS LATER, however, when I saw Birbaumer, he seemed to have lost some of his confidence in the experiment. "We're not moving," he said. "Well, we're moving, but we're not moving fast. To make substantial progress, which I think we could do, I would need about ten times the money." In part, his problem lies with the success of the "monkey guys"—his term of half-affectionate disregard for researchers on the invasive side of BCI, those who work with implanted electrodes that pick up the brain's actual electrical instructions to the body, rather than reading, at the scalp, the byproduct of those instructions.

At a BCI conference held in 2002 in Rensselaerville, New York, a researcher of the invasive school was overheard comparing the difference between his research and noninvasive research to jet airplanes and hot-air balloons. A balloon was inarguably colorful and charming, and it worked, but to reach a conference one might prefer a 747. Animal experimenters at Duke University and at Brown University, among other places, are perhaps closer to science's cutting edge than Birbaumer is, and they are more ready to think beyond communication devices to systems of brain-controlled motion. Some have formed private companies, and some receive defense funding. ("What the military really wants is a remote-controlled soldier," Birbaumer says.)

Birbaumer takes a kind of moral strength from working with people—a dozen so far—who have immediate communication needs, rather than working with animals, but he is happy to concede that the monkey researchers have begun to produce astonishing results. They have found, for instance, that one can know the intention of a brain by keeping track of surprisingly few of its billions of neurons. At Duke University, the Brazilian-born neuroscientist Miguel Nicolelis, who has some of Birbaumer's flamboyant style—his Web site plays "Bim Bom," "Girl from Ipanema," and other Joao Gilberto hits—used techniques developed with John Chapin, now at the State University of New York Health Science Center, to implant hundreds of Teflon-coated microwires, bundled together in groups, into the brains of two owl monkeys, in areas of the cortex known to be associated with motor function. A computer algorithm combined the information from the electrodes. Nicolelis trained the monkeys to make particular arm movements, and, over two years, he read and recorded patterns of neuronal firing associated with those movements. He was identifying the neuronal signature of a particular monkey gesture. In 2001, he was able to use a signal taken directly from the animal's cortex to make a robotic arm simultaneously replicate the monkey's actual arm movement. (And, for good measure, the signal was sent across the Internet, to make a robotic arm move at MIT, six hundred miles away.)

In 2002, at Brown University, a team lead by John Donoghue, the chairman of the Department of Neuroscience, reported something similar: a monkey first learned to use a joystick to play a video game, while his keepers similarly built up a key—a code book—linking neuronal activity with various actions. Then the joystick was disconnected from the computer. The monkey continued to play the game using only his brain.

The clearer signals extracted from the brain in these experiments, and others similar to them, may turn out to be suited to the control of wheelchairs or limbs—or even devices of "neurological augmentation" for able-bodied people: speed typing, for example. "Exciting things are going to happen in the next couple of years," Donoghue told me. "We're trying to re-create the nervous system, and I've been shocked how easy it is. You know, it almost seems too good to be true. You have these nights—'How is it so easy to get these ideas out of the brain from such small numbers of neurons?' "

Barry Dworkin, a leading neuroscientist and psychologist at Penn State's Hershey Medical Center, says, "There's a technical difference, not a categorical difference, between what Nicolelis and Birbaumer are doing. This is all about information—about how much information you can extract. No doubt the theoretical capacity of electrodes surgically implanted in the brain is vastly greater than Birbaumer's slow potentials, but the electrodes are highly invasive. And it is actually quite hard to extract clean signals from them, and to transform them into real-time movement schemes." He adds, "Our natural speech and motor apparatus works so well it's hard to see the applications for able-bodied people, even with a vastly improved technology. I suppose you could wire people's brains together instead of making them meet in a committee room. That might save time. At the moment, for disabled people, there's a lot to be said for Birbaumer's method, if it works. It's not invasive and these people are pretty fragile. Of course, you'll never be able to use it to fly an F-16."

THE INVASIVE-TECHNIQUE SCIENTISTS may have drawn attention and research grants away from Birbaumer, but something else was feeding into his restlessness last summer. It was the nagging possibility that, whatever else he would achieve with the TTD, he might never use it to communicate with people fully locked in their bodies. The system has not been proved in the environment where it is most needed. Hans-Peter Salzmann still uses his facial muscles for everyday talk: BCI is a secondary tool, for private letters and e-mails generated over weeks and months—a pocket of autonomy in a life that is otherwise entirely collaborative.

Meanwhile, Birbaumer's few fully locked-in subjects—those who have no residual facial movement—remained silent. Among them was a woman paralyzed by Guillain-Barré syndrome, a rare neurological condition. A coma test showed that she was "still with us," Birbaumer said—her EEG showed that she took notice, for example, of auditory stimuli—and she seemed at first to be responsive to training, but "now we've seen nothing for more than a year." A patient with hereditary ALS had begun training well before he became locked in, but then his wife left him, and he was moved into a nursing home. Now fully locked in, he rarely has visitors; his son, who has a fifty-per-cent risk of inheriting the disease, finds it hard to be with him. The patient has stopped responding. "He doesn't want to communicate," Birbaumer said. "He doesn't answer me. I asked him if he wanted to live or not to live. I said, 'You can tell me.' I always say to my patients, 'If you don't want to live, and you have good reasons, I will help you.' But he never said that."

Was it possible that the loss of one's last faint muscle movement somehow eliminated the will to be heard? Birbaumer instinctively resists this idea. He was always upbeat about the potential quality of life of people who are unable to move, and contemptuous of the assumption made by some neurologists that when a patient decides to live beyond the point of artificial respiration he or she enters a zone of hopeless despair. Birbaumer's reading of the literature, and his own research, has suggested to him something more positive. A study done in his department compared the mental life of eighty ALS patients at various stages of the disease, eighty people who had been given a diagnosis of clinical depression, and a sample of those whom Birbaumer calls, with a touch of comic disparagement, "normals." "We found that the ALS patients are significantly more depressed than the normals, but they're significantly less depressed than the depressed; and they're all within the normal range." So, in his opinion, the usual neurological assumption is "just plain bullshit," he said.

But this leaves aside a more focussed question about the mental life of those few people who are, like Elías Musiris, completely locked in. "I was once much more secure in the idea that these people really think normally and their inner life is intact," Birbaumer said. "Now I'm not sure."

He had to leave to give a lecture, but, before he went, he said, "You can make yourself die. There are data on that. You can shut down your organism and kill yourself by just doing nothing. In hopeless situations, the immune system falls apart, then you die of an opportunistic disease. You give up."

THE ATLANTIC CITY CASINO occupies the first three floors of a gray office building at a key intersection in the prosperous but not pretty Lima neighborhood of Miraflores. When I visited late one night in July, most of the customers were well-dressed middle-aged women; the slot machines produced a dense symphonic din. On a second-floor mezzanine, a little bar had been built underneath a staircase, and the wall behind it was studded with pinpoints of light to resemble a night sky. Elías Musiris was sitting in the dim light, his wheelchair facing a TV monitor that showed security-camera images of the casino, just as in his dining room at home. Estrella was with him. A heartbeat monitor clipped to his earlobe glowed green.

As a healthy man, Musiris spent every night at his casino. He still arrives every evening before midnight, and leaves every morning at dawn. Then he sleeps—the only sign of this is that his eyes move back a little in their sockets—and a nurse watches his face constantly, scanning for signs of crisis. Sometime after midday, he is put into his wheelchair and taken to the dining room. Twice a week, he is washed, and he has a daily massage. A doctor is always on hand. Polo asks questions using his pulse system, and the nurses read to him from the morning newspaper. Then, at night, he is driven again to the casino. "He said he wanted to go to the casino, and for everything to stay the same," Estrella said, when I sat with them. A friend of Musiris's—a fellow textile business-man—came by and kissed him on the forehead. Polo dashed about, now and then changing the image on Musiris's TV screen. I watched him zoom in tightly on a machine where casino guests were making bets on a computer-generated horse race. Bright animated horses filled Musiris's screen: a silent, security-camera image of a cartoon representation of his favorite sport.

Birbaumer's hotel was directly across the street from the casino. The morn-ing after his first session with Musiris, Birbaumer was sitting in the hotel lobby with José Luis Palomino. Birbaumer describes himself as a behaviorist—enjoy-ing the slight datedness of the label—and he makes respectful reference to the work of the late B. F. Skinner, who trained pigeons to peck at a colored plastic disk by rewarding, first, any tiny accidental movement toward the disk, then any closer movement, then actual contact. Skinner called it "shaping"—where a desired form of behavior, even one well outside an animal's usual repertoire, can be fashioned out of a series of small steps taken in the right direction.

"The important part is going to be psychological," Birbaumer said. "The reward. We need maximum reward. Elías is a thinking, feeling person." He glanced across the street. "Can you imagine someone saying, 'If you win on the slot machine, we'll give you the money tomorrow'? We can't just say, 'You will communicate in the future.' That's not enough." (Birbaumer lobbied in his university department for using "sexual reinforcers"—sexual imagery or visits from prostitutes—as a form of "intermittent rewarding stimulus" for some of his patients, who are mostly male. He was voted down by colleagues.)

He and Palomino did not doubt that Musiris had understood them. In a passive "auditory oddball" test, a series of beeps on one note had been played to Musiris, then a beep on another, lower, note. This was repeated. According to his EEG reading, he had registered the different beep, repeatedly, meaning that he had short-term memory and some ability to notice changes in his environ-ment—which is not always true of ALS patients. Another test was also positive.

For Birbaumer and Palomino, then, the issue was will, not skill. How could Musiris be motivated? How could his behavior be "shaped"? "Could we try this?" Palomino asked. He clapped his hands. "Applause. He's an Arab. He likes to have people around him."

"The problem is, human reinforcement might be distracting," said Bir-baumer, who had disliked how busy the room had been the previous day. "I was thinking, What's important to him? And it's maybe business. With Salzmann, his control is over us, and the material he keeps secret from us. That's what keeps him alive. We have to think of something that Musiris can take control over."

BACK AT THE HOUSE, Musiris was brought into the poolside room and the electrodes were again fixed to his head. *"En sus marcas, listos, ya!"* Palomino said: ready, set, go. The ball of light moved across the screen. To everyone's surprise, Musiris scored a string of hits. The ball seemed to have a new lightness. Musiris managed sixteen hits out of twenty, and the room burst into applause. Polo punched the air with his fist.

But in a trial a few minutes later Musiris scored only seven out of twenty. The following evening, this pattern was repeated: early in the session, he did well, and for a moment he seemed to reveal himself—he had a new, fuller presence in the room. But then he faded, pulled away from his audience by low scores.

Birbaumer had brought a videotape from Germany on which Salzmann was seen slowly spelling out "ELÍAS, U CAN DO IT" with the TTD—but Birbaumer did not quite share Salzmann's confidence. Musiris was responding better than any other fully locked-in patient Birbaumer had seen. But until Musiris could consistently move the light into the correct half of the screen with at least seventy-five-per-cent accuracy, there would be little point promoting him to the letter-selection program used by Salzmann: the error rate would undermine the system and perhaps crush Musiris's morale.

Daily practice would surely improve the patient's average, but Birbaumer would be in Lima for only three weeks. So while Musiris continued to train in the usual way, with the eventual goal of creating words, Birbaumer and his colleagues began to improvise a simpler system inspired by work he had done years before on lie detection. ("I hated it for ethical reasons, but it was interesting scientifically," he told me.) Musiris would be read a question or a statement. If he sent the ball up, he would be responding "yes"; if he sent the ball down, he would be saying "no." This would be reinforced by a recorded voice saying *"sí, sí, sí, sí"* when the ball was above the line, and *"no, no, no, no"* when below. (Birbaumer was increasingly persuaded that Musiris was not seeing clearly, if at all.) As with lie detectors, Musiris would be asked the same thing repeatedly, in both a positive and a negative form: both "I want to go to the casino" and "I don't want to go to the casino." These would be intermixed with control sentences to which the correct response was known: "Lima is the capital of Peru";

"Lima is not the capital of Peru." One could, in theory, spell this way, at an absurdly slow pace. But the system's value was as a form of actual, if crude, communication, and therefore an encouragement. A simple yes-no system would forgive a very modest hit rate. Even a faint bias would become statistically persuasive—statistically unassailable, eventually—given enough repetition. "We can tell his intent from sixty per cent," Birbaumer said. If a tossed coin lands heads up six times out of ten just once, that says nothing; if it happens for a week, then something has been done to the coin—a bias has been proved.

Birbaumer badgered the family for questions that would engage Musiris. Birbaumer also wrote his own list. Although he was keen to ask a question about Elías's feelings for Estrella, he could not see how to express the negative without causing offense: "I love my wife. I hate my wife." "We have to mix the psychology and the mathematics," Birbaumer said. This process of fusion was wearying: he and Herta Flor were also trying to adjust the system to reward even ambiguous responses—and to allow for the fact that Musiris found it easier to take the ball up than down. This put them in frequent conflict with Thilo Hinterberger, who did not want to introduce imprecision into his computer program. In daily arguments over this question, Hinterberger turned from cool calm to red-faced fury in seconds, shouting, "I know what I'm doing!"

For all the conflict, Birbaumer was also stimulated. "I don't usually follow these patients so closely," he said. "You know, I do the business, I get the money, I kick the young people"—his research students—"in the ass, that's my job. I never sit here like this. I don't have time. We should do it like this always. That's the right way to proceed in science. But it's not what's allowed by the system."

On Monday, Birbaumer's fourth day in Lima, they started asking Musiris the yes-or-no questions. "Do you like business?" a recorded voice said. Musiris replied, "*Sí, sí, sí, no, no, sí, sí, sí*"—the ball mostly stayed above the line. "Do you not like business?" "*No, sí, no, no, no, no, no.*" The question was asked six times, three times in each form. He gave the same answer five times. Then he was asked another six times: now he liked business three times. Then: "Do you like racehorses? Do you not like racehorses?" "Is your manager Eddie Thornberry?" "Do you like going to the casino?" "Should the casino have two new roulette tables?" At Birbaumer's suggestion, Estrella had taken a place by her husband's side, from where she was saying "*Muy bien, muy bien,*" and blowing him kisses.

Despite the reinforcement—the kisses, the "*sí, sí, sí, sí,*" the promise of immediate communication—Musiris's percentages did not show a dramatic rise, and again he was being dragged into the shadows of poor percentages, perhaps by tiredness. But with the new system repetition could squeeze meaning out of a hint. The casino questions, in particular, had begun to generate fairly reliable answers. If one accepted the math, then Musiris did want to go to the casino; and they should buy the new tables.

Musiris left for his night at the Atlantic City. Had there been communication? "We think so, but it's very rudimentary," Birbaumer said. He had seen a breakthrough, of sorts—a dialogue with a fully locked-in patient—but he could not be described as upbeat. "We're getting probabilities, but it's not what people like. People don't like statistics." He frowned and shrugged.

Every evening, Musiris came to the den to train. After a week, the yes-or-no system had drawn answers from him that, through repetition, had reached a statistical safety level of more than ninety per cent. Some questions about his domestic arrangements never seemed to engage him: "Do you want to sleep?" "Do you want to stay in this room?" But questions about business reached a level of statistical certainty. "Should we invest in a new textile-dyeing system?" Thornberry asked him. Musiris said "yes," emphatically. "May we sell the old slot machines?" Thornberry wanted to sell them, but Musiris, who was asked more than fifty times, said "no" repeatedly. Thornberry said he would respect that decision. Musiris was beginning to have a deliberate impact on the world without moving a muscle.

He continued to train toward the goal of letter selection. At the end of the second week, Birbaumer presented him with a version of Salzmann's spelling program where the alphabet had been reduced to a handful of letters. Birbaumer gave him a spelling task—two words to copy—but Musiris struggled. At one o'clock in the morning, he had produced nothing. Birbaumer came close to him, and said, "If you spell this correctly now, then you can go to the casino." Over the next twenty minutes, without making an error, Musiris selected "E," "L," "I"—until he had spelled out "ELÍAS MUSIRIS." His family shouted, danced, and kissed one another. "He wrote his own name. That was incredible," Birbaumer later remembered. "The family went crazy."

I TELEPHONED BIRBAUMER a few weeks after he had flown home to Tübingen. "I'm pessimistic about getting him to do real free spelling, using the whole alphabet, and choosing his own words," he said. "But we did communicate with him, and we still are communicating. And I'm more convinced than I was before that the TTD can be useful for people who are completely locked in." Birbaumer was thinking about future publication. Musiris was continuing to train with Palomino, who was e-mailing data to Germany each day, although the experiment had been interrupted for several weeks when Lorenzo, Estrella's pet monkey, damaged the amplifier. Lorenzo was now living with neighbors.

While newly optimistic about the possibility of some form of communication with fully locked-in people, Birbaumer had revised his sense of the mental lives of these patients. "I have no data on this—it's completely subjective—but now I think that when you become locked in you also partially lock yourself in cognitively," he said. Hans-Peter Salzmann will avoid it, perhaps, by training throughout the period during which he becomes fully locked in—if that is his fate. But Musiris seemed partially marooned. "Maybe after a while you stop thinking in particular areas," Birbaumer said. "Probably a lot of your inner life becomes empty, and you have some residual modules in your brain. In Elías's case, let's call them the casino module, the factory module, the Estrella module, which can be activated from time to time, but most of the time you're in a locked-in state mentally—there's nothing going on in your head anymore. That was my impression with him. In most situations he had just stopped thinking. But this is nothing more than a clinical impression. And you know what that's worth—nothing." He laughed.

Birbaumer was once able to think of his work as an act of subversion; he would introduce into public discourse the voice of a group silenced by neurologists, and they would deliver reports from an unknown place of contemplation where they had been loosened from social norms. "I would have caused a revolution!" he said. "I would have liked that when I was young. I always liked revolutions." The truth was perhaps quieter. "In Elías's situation, I can imagine myself sitting in the chair and thinking a lot, but also not thinking for long periods of time, and maybe communicating with this kind of instrument just now and again," he said. "Rather than our blah blah blah, which we do day and night, which is no use to us. And you see that life comes down to fifty questions! We were wringing our hands—'What can we ask him, what can we ask?' Because in the end what are the crucial questions?" Even in the case of Elías Musiris, a patient with a family, a fortune, and a business empire, there seemed at times to be nothing to say, and no overpowering need for information to pass in either direction. "You ask them, 'Do you want to live? Do you want to die?' After that question, everything else is kind of trivial."

TOM SIEGFRIED

The Science of Strategy

FROM *THE DALLAS MORNING NEWS*

M illions of people have seen the movie, and thousands have read the book, but few fully grasp the math invented by John Nash's beautiful mind.

Nash, the troubled mathematician played on film by Russell Crowe, won the 1994 Nobel Prize in economics for pioneering research in the 1950s on an arcane branch of math called game theory.

For decades, few researchers outside economics paid much attention to game theory. But in the 1980s, some biologists began having fun with it, and now scientists from other disciplines are learning how to play as well.

Offshoots from Nash's work have sprouted in several scientific fields, entangling economics with biology and psychology, sociology, neuroscience and even quantum physics. Nowadays game theory turns up in everything from fighting terrorists to designing smarter robots. And game-theory-based studies of the brain have challenged basic tenets of economic science, revealing surprising nuances in human behavior.

"I think in 10 years, economics will be completely changed because of this work," says Paul Zak, an economist at the Claremont Graduate University in California. "Literally, the textbooks will be rewritten."

At the time of Nash's early work, game theory was briefly popular among

Cold War analysts as well as economists studying how people make profit-maximizing choices.

"But it didn't take off," says Sam Bowles, an economist at the University of Massachusetts Amherst. "Like a lot of good ideas in economics, it just fell by the wayside."

By the early 1990s, though, game theory had recaptured economists' attention, as recognized with the Nobel award to Nash and two other pioneers in the field, John Harsanyi and Reinhard Selten. Even before then, though, biologists had begun to exploit game theory's insights, applying them to the competition for survival among animals and plants. More recently psychologists have used game theory to probe the inner workings of the human mind. And with the turn of the new century, neuroscientists have joined in the fun, peering inside the brains of game-playing people to discover how their strategies reflect different motives and emotions.

"We're quantifying human experience in the same way we quantify air flow over the wings of a Boeing 777," says neuroscientist Read Montague of Baylor College of Medicine in Houston.

Game theory's roots go back centuries, but the first explicit presentation of the math came in a 1928 paper from the Hungarian mathematician John von Neumann. Building on that work, von Neumann collaborated with the economist Oskar Morgenstern to produce the bible of the field, *Theory of Games and Economic Behavior,* in 1944.

"We hope to establish," they wrote, "that the typical problems of economic behavior become strictly identical with the mathematical notions of suitable games of strategy."

But game theory's original formulation was limited, focusing mostly on two-party "zero-sum" games in which one player wins only what the other player loses. Nash expanded the scope of game theory to more realistic economic situations.

In one paper, he analyzed the problems faced by two negotiators attempting to reach an agreement—or make a bargain—that maximizes the payoffs for both parties. In other work, he treated non-zero-sum games, where one player's winnings do not have to equal another's losses. Nash established that for many types of such games, with multiple non-cooperating players, some best-case scenario for all concerned can be computed using game theory. In other words, all interests can be balanced in what is now called a "Nash equilibrium."

"The concept of the Nash equilibrium is probably the single most fundamental concept in game theory," says Dr. Bowles, who is also affiliated with the Santa Fe Institute in New Mexico.

Nash's results hinged on assumptions central to economic theory, especially the notion that people behave rationally and act in their own self-interest. Put another way, the No. 1 rule is that everybody is out for No. 1. Game theory provided a way to quantify that notion in terms of "utility," the value a rational agent assigns to goods or actions when making economic decisions.

GIVEN THE ADDITIONAL ASSUMPTION that you have enough information to know what's best for you, game theory provides straightforward math for choosing the optimum economic strategy. In principle, such math could guide negotiations between labor unions and management, help nations reach trade agreements or improve the prospects of winning in Vegas. (Von Neumann's work relied heavily on analyzing common card games like poker and bridge.)

In practice, though, people don't always choose the strategies that standard economic theory predicts. Part of the problem, says Dr. Bowles, is the old textbook view that people in an economy interact only with prices. It's known as the Robinson Crusoe model—living on an island, he interacted only with nature.

But in a real economy, nobody's an island; people interact with other people, making choices not only in response to prices but also by anticipating what the other guy will do next, an element missing from the Robinson Crusoe approach.

"Game theory adopts a different framework," Dr. Bowles said in a telephone interview. "My well-being depends on what somebody else does, and your well-being depends on what I do. Therefore we are going to think strategically."

Choosing sound strategies in such interactions is what game theory is all about. It provides a method of testing how people's actual strategies compare with what standard economic theory predicts.

"GAME THEORY has really provided a way of posing the question . . . in a very clear way," Dr. Bowles said. "Game theory is also then part of how you solve the problem."

Disenchanted economists began challenging standard theory in the 1980s by devising laboratory experiments to gather data on how people really do reason and choose. (George Mason University economist Vernon L. Smith shared the 2002 Nobel in economics for early work in that field.) Data from such experiments fueled the development of a field known as behavioral economics.

"Behavioral economics is a school of thought that says let's go back and question some of the basic mathematical foundations of economics," says Colin Camerer, a professor of business economics at the California Institute of Technology in Pasadena.

Dr. Camerer specializes in the branch of behavioral economics called behavioral game theory, using games to test how psychological realities drive economic behavior that departs from standard assumptions.

After all, he notes, there is nothing sacred about those assumptions. Many of them originally arose mainly for computational convenience.

Most economists have known all along, for example, that people are not always rational and seldom have complete information. But early efforts to make economics rigorous had to ignore such complications. There are many ways, for instance, in which people can behave irrationally. Assuming rational behavior was the only way to be mathematically precise, many economists maintained.

"They've often used that as an excuse," Dr. Camerer said in an interview. "But our view is completely different, which is to say let's ask scientists who have been thinking about how brains actually work . . . for some help."

There may well be many ways to be irrational from an economist's point of view, he said. But psychologists may be able to discover in which ways people actually are irrational.

"Economists have been kind of conservative about this, saying if you give up rationality, we'll never be able to have anything precise," said Dr. Camerer. "But we will, and it's emerging pretty rapidly by using the evidence."

An important new source of such evidence is the newborn discipline of neuroeconomics, where researchers can spy on people's brains during economic game-playing.

MIXING ECONOMICS with neuroscience is the ultimate no-brainer. After all, says Dr. Zak, economics is about making decisions and choosing behaviors, actions governed by the activity of nerve cells in the brain.

"There's such affinity between these two groups, it's absolutely natural to combine them," he said in an interview.

Game theory on its own can't answer questions about why people make certain choices, he points out. But experiments combining games with brain scans can reveal the players' underlying mental activity responsible for the behavior.

"If you ask subjects, why did you make this decision, they can't tell you," Dr. Zak said. "Using neuroscientific techniques allows us to get around that problem. I can just measure directly what's happening."

Neuroeconomics is far from the only realm where game theory has gained in popularity. In battling terrorism, for example, it has emerged as an important way to evaluate standard approaches to analyzing intelligence data and formulating policy.

Todd Sandler, a political economist at the University of Southern California, has pursued game theory's implications for terrorism since the 1980s. Papers before then had discussed various aspects of anti-terrorism policy, but none with game theory's mathematical rigor.

"There was no theoretical framework to looking at how government decisions may interact with terrorist decisions," Dr. Sandler said in an interview at USC. "And that's what game theory does, it looks at the interactions between the two players. . . . It often gives results that are not intuitively obvious at first."

Another possible application of game theory to terrorism comes from studies of smallpox vaccination strategies. Chris Bauch and David Earn, of McMaster University in Canada, and Alison Galvani, of the University of California, Berkeley, have combined game theory with the math describing epidemics to assess the policy of voluntary smallpox vaccinations. Individual decisions guided by self-interest are likely to produce a less-than-optimum level of vaccination for the whole population, the researchers reported in September 2003 in the *Proceedings of the National Academy of Sciences*.

Game theory's ramifications extend even to the physical sciences and technology. For example, game theory has guided experiments for teaching robots how to communicate. Robots programmed to play language games can teach themselves new words to describe objects in their environment. Sounds correctly linked with meanings earn the robots higher scores in the games, as described by Luc Steels of the University of Brussels and the Sony Computer Science Laboratory in Paris in a recent issue of *Trends in Cognitive Sciences*.

AMONG PHYSICISTS, interest is growing in quantum versions of game theory, in which the weird math of the subatomic world offers new twists to standard game strategies. In quantum physics, many possibilities can exist simultaneously. Adapting quantum math to decision-making in games provides a richer repertoire of choices than mere yes-or-no responses.

David Meyer, a mathematician at the University of California, San Diego, introduced a simple quantum game in a 1999 paper published in *Physical Review Letters*. Progress in devising more elaborate quantum games has so far been mixed, he said in an interview. But communication among game players using quantum signals is technologically possible, he points out. Combined with other new online communications techniques, quantum information could give birth to new species of games.

"The Internet and modern telecommunications are likely to change the way we think about game theory," he said.

Even without quantum leaps, though, game theory is likely to assume a more prominent role in economic policy, especially in the wake of advances in behavioral game theory and neuroeconomics.

"We're not quite ready to make too many policy prescriptions," said Dr. Camerer, "but people have a few ideas."

Ultimately, he believes, policy-makers will recognize that the game-behavior approach to economics is just a way of making it more realistic, unconstrained by the old ideas about people being perfectly rational.

"Behavioral economics," says Dr. Camerer, "is kind of a step back to something which is more sensible."

KAJA PERINA

Cracking the Harvard X-Files

FROM *PSYCHOLOGY TODAY*

People who believe they've been abducted by aliens have always resided at the farthest fringes of science, and the recent claim by a UFO cult known as the Raelians that they had cloned a human being does little to endear abductees to the mainstream. The sect's leader, Rael, maintains that he was plucked from a volcano by almond-eyed aliens who granted him an audience with Jesus, Buddha and Muhammad, each of whom confirmed that humans are descended from extraterrestrials.

But for every Rael, there are hundreds of workaday individuals who claim to have been abducted by aliens. These individuals do not flower into gurus; they struggle alone with memories of unintelligible messages, temporary paralysis and humanoid creatures hovering over their beds. Their stories don't always check out, but their minds do: Psychological tests confirm that abductees are rarely psychotic or mentally ill. Some 3 million Americans believe they've encountered bright lights and incurred strange bodily marks indicative of a possible encounter with aliens, according to a recent poll.

It is a quandary that polarizes researchers at Harvard University. One embattled psychiatrist, John Mack, MD, argues that these experiences cannot be understood in a Western rationalist tradition of science; researchers in the department of psychology, Richard McNally, PhD, and Susan Clancy, PhD, counter that the explanation—though multifaceted—is hilarious in its fundamental simplicity.

Mack, of Harvard Medical School, is a long-time champion of alien abductees and a paranormal philosopher king of sorts. His 1994 bestseller, *Abduction: Human Encounters with Aliens,* drew international attention with the argument that "experiencers," Mack's term for the men and women he has debriefed, probably are being abducted by aliens.

More recently, McNally and Clancy introduced alien abductees to the laboratory to study trauma and recovered memory in an experimental setting. They believe their subsequent findings explain the entire abduction experience, including abductees' refusal to accept the fact that transcendent, technicolor encounters with aliens are no more than five-alarm fires in the brain.

Harvard's ideological clashes over the interpretation of anomalous experiences date to William James's tenure at the university one century ago. Both Mack and James studied psychology after training in medicine and tried to bridge the gap between psychology and spirituality, only to be rebuffed by Harvard's powers that be. For James, this culminated in *The Varieties of Religious Experience,* which rejected a rigorous standard of evidence for divine experiences. "There is a clinical literature and an experimental literature, and they don't refer to each other," states Eugene Taylor, PhD, a biographer of James and a historian who lectures on psychology at Harvard Medical School. "Mack is a clinician making observations about human experience, as opposed to cognitive behavioral scientists, who say that if you can't measure it in the laboratory, it doesn't exist." When it comes to people who believe they've been abducted by space aliens, the two camps agree on only one thing: "These people are almost never psychotic," says McNally. "They're not lying. But Mack entertains a range of explanations that are farfetched at best."

WILL BUECHE, a 34-year-old media director, has long had nighttime paralysis and visions that "have no resolution and seem out of place." For years, he considered them merely suggestive—until he began witnessing beings while wide awake. Some abductees had far more traumatic encounters. Peter Faust, a 45-year-old acupuncturist, believes he endured years of sexual probing by hooded creatures who implanted chips in his anus and stimulated him to ejaculation. After eight hypnotic-regression sessions with Mack, and a battery of psychological tests in the early 1990s, Faust concluded that he is yoked to a female alien-human hybrid with whom he has multiple offspring.

The abduction narrative is a strange hybrid in its own right: humiliating surgical invasion tempered by cosmic awareness. Experiencers travel through windows and walls, tunnels and space-time to reach the starship's examining table, where young women's eggs are extracted and men's sperm are siphoned off. Despite waking bruised and violated, abductees say their love for beings in the alien realm can surpass any human bond and generate a sense of oceanic oneness with the universe that rivals the experiences of a world-class meditator. Faust says he "realized we're not alone in the universe. There are beings out there who care about us. But getting to this point is a long, arduous journey, with a lot of people who want to deny your experience."

Personality-driven explanations for why people with no overt psychopathology report alien encounters have proliferated apace with blockbuster movies about aliens. Psychologist Roy Baumeister, PhD, of Case Western Reserve University, argues that abduction reports are made by "masochists" who unconsciously want to relinquish control of their lives. The loss of control is manifest in humiliating encounters with an alien race. To be sure, there is a surfeit of elaborate sex in abduction reports; one study found that among abductees, 80 percent of women and 50 percent of men reported being examined naked on a table by humanoid beings. In fact, many abductees blame aliens for sexual dysfunction and emotional disturbances.

Psychologists have long surmised that abductees may be inclined to fantasy and "absorption," the propensity to daydream or be enthralled by novels. Both alien abductees and garden-variety fantasizers report false pregnancies, out-of-body experiences and apparition sightings. Some psychologists speculate that people like Will Bueche and Peter Faust are simply "encounter-prone" individuals with a heightened receptivity to anomalous experience. Whatever the case, Bueche and Faust found a willing listener in John Mack.

MACK HAS BEEN on the faculty of Harvard Medical School since 1955, and in 1982 he founded the Center for Psychology and Social Change, located in a yellow clapboard house just beyond the university's campus. The Center aims in part to study anomalous experiences, and has its post office box in Cambridge, but the building lies just within neighboring Somerville. The address is a fitting line of demarcation for a clinician who straddled conventional science and altered states of consciousness long before the publication of *Abduction*.

Mack founded the department of psychiatry at The Cambridge Hospital in 1969; a program that has long attracted innovative, Eastern-oriented psychiatrists. In 1977, Mack was awarded a Pulitzer Prize for *A Prince of Our Disorder*, a biography of Lawrence of Arabia. "Mack is in dynamic communication with the humanities," says Eugene Taylor.

Mack has embraced traditions from Freudian psychoanalysis to the guided meditation of Werner Erhard. In 1988, he began to practice Stanislav Grof's holotropic breathwork, a technique that induces an altered state by means of deep, rapid breathing and evocative music. Mack believes he retrieved memories of his mother's death, which occurred when he was 8 months old. "I was raised in a tradition of inquiry," says Mack. "If you encounter something that doesn't fit your worldview, it's more intellectually honest to say, 'maybe there's something wrong with this worldview,' than to try to shoehorn your findings into an existing belief."

At 73, Mack appears regal despite his slightly stooped gait. His handsome, deeply lined face and flinty blue eyes are quietly compelling; he quickly earned a reputation for emotional succor among the abductees he interviewed. Abductees including Faust and Bueche cling to him like acolytes, often parroting his theories.

Mack used hypnotic regression to retrieve detailed memories of 13 encounters with aliens, all chronicled in *Abduction*. He has now interviewed more than 200 abductees. He says that he ultimately endorsed abduction reports largely because he found his subjects to be mentally competent. Some were also highly traumatized and most were reluctant to come forward and appropriately skeptical about their experiences.

Mack defends the use of controversial techniques such as hypnotic regression because he prizes the experiential narrative over empirical data. To debrief an abductee is to be "in the presence of a truth teller, a witness to a compelling, often sacred, reality." Mack says he was jolted when his subjects reported receiving telepathic warnings about man's decimation of natural resources. "I thought this was about aliens taking eggs and sperm and traumatizing people," admits Mack. "I was surprised to find it was an informational thing."

The faculty of Harvard Medical School, for its part, was dumbfounded that Mack believed he'd stumbled on anything more than an underreported cluster of psychiatric symptoms. From 1994 to 1995, Arnold Relman, MD, professor emeritus of medicine, chaired an ad-hoc committee that conducted a 15-month investigation into Mack's work with abductees. "John did good things in his career and gained a lot of respect. His behavior with regard to the alien-abduction story disappointed a lot of his colleagues," says Relman. The investigation ended with much tongue-wagging but no formal censure. Mack was, however, encouraged to bring a multidisciplinary approach to his study of the phenomenon. "No one is challenging John's right to look into the matter," sighs Relman. "All we're saying is, if you do it, do it in an objective, scholarly manner."

In the spring of 1999, Mack invited astrophysicists, anthropologists and a Jungian analyst who studies anomalous experience in the wake of organ transplants to the Harvard Divinity School, where they brainstormed with mental health professionals and abductees. One participant was Harvard psychology professor Richard McNally, an expert on cognitive processing in anxiety disorders.

McNally told the assembly that "sleep-related aspects of the experiences might be correlated with different parts of the REM cycle." He was referring to the phenomenon of sleep paralysis, but he hesitated to speak bluntly about it. Many abductees deem sleep paralysis too mundane an explanation for their experiences, so McNally didn't use the term, for fear of "alienating" the very subjects he wanted to recruit.

SLEEP PARALYSIS IS a common phenomenon—up to 60 percent of people have at least one episode, in which the brain and body momentarily desynchronize when waking from REM sleep. The body remains paralyzed, as is standard during the REM cycle, but the mind is semi-lucid or fully cognizant of its surroundings, even, according to a Japanese study, if one's eyes are closed. The experience can't be technically classified as either waking or sleeping. For an unlucky handful of people, fleeting paralysis is accompanied by horrifying visual and auditory hallucinations: bright lights, a sense of choking and the conviction that an intruder is present. The Japanese call it kanashibari, represented as a devil stepping on a hapless sleeper's chest; the Chinese refer to it as gui ya, or ghost pressure.

Sleep paralysis with hypnopompic hallucinations (those that occur upon waking) can be so unexpected and terrifying that people routinely believe they're stricken with a grave neurological illness or that they're going insane. When faced with these prospects, aliens no longer seem so nefarious.

But sleep paralysis and abduction don't always go hand in hand. Consider the case of "Janet," a 52-year-old copy editor in Chicago. Eleven years ago she endured a terrifying out-of-body experience while lying in bed. Janet saw her head strapped in a vise as a group of men looked on. Fuzzy images were projected onto the back of Janet's eyes, visions she likens to "a 3-D hologram engraving something into my head." Her first thought on waking was of a brutal sexual assault she'd once read about. McNally believes it is the sense of powerlessness in being immobilized that triggers associations with invasive sexual procedures.

Janet experienced terror and helplessness in the wake of these messages she could not decipher, and sought the help of numerous therapists. But she says she "never thought this had anything to do with aliens. I thought it was something arising from the depths of my subconscious."

Why, then, do some people who experience violent hallucinations upon waking or falling asleep conclude that they have been abducted? One possibility is that people embellish their experience in the course of hypnotic regression. But McNally and Susan Clancy speculate that alien abductees aren't just amenable to suggestion under hypnosis; instead they actively create false memories. They drew this conclusion while studying one of the most contentious issues in psychology today: false memory syndrome.

THE QUESTION OF whether or not people repress traumatic memories was thrown into high relief 15 years ago, as psychotherapy patients increasingly recovered memories of sexual abuse, often through such porous techniques as hypnotic regression and guided imagery. Some cognitive psychologists, including McNally, argued that people rarely repress memories of abuse or trauma; if anything, they are more likely to recall the incident. Sexual-abuse victims remain silent "not because they are incapable of remembering, but because it's a terrible secret," says McNally. Other professionals argue that traumatic memories are easily repressed through specific dissociative mechanisms.

In 1996, McNally and Clancy became the first researchers to examine memory function in women who believed they had recovered memories of childhood sexual abuse. They found that these women were significantly more likely to create false memories of nontraumatic events in a lab than were women who had always remembered being sexually abused, or women who had never been abused. (The findings are outlined in McNally's book, *Remembering Trauma*, published in the spring of 2003.)

False memory was assessed by asking subjects to study semantically related words (such as *candy, sugar, brownie* and *cookie)* and then identify them on a list that includes false targets such as *sweet*, words that are thematically similar but not previously presented. Members of the recovered-memory group were by far the most likely to believe they'd seen the false targets.

But McNally and Clancy could not ascertain whether the women had in fact been sexually abused. Since it is unethical to create false memories of trauma, the researchers did the next best thing: They amassed a group whose recovered memories were unlikely to have occurred. Those people were, of course, alien abductees.

McNally and Clancy assembled a group whose members believed they'd recovered memories (usually under hypnosis) of alien abduction, along with a repressed memory group whose members believed they'd been abducted but had no conscious memory of the event. (This group inferred their abduction from physical abrasions, waking in strange positions or sometimes just from their penchant for science fiction.) There was also a terrestrially bound control group who reported no abduction experiences.

The recovered and repressed memory groups exhibited high rates of false recall on the word-recognition test. Those with "intact" memories of abduction fared worse than those who believed their memories were repressed.

But could this type of false recall be a function of memory deficits incurred through traumatic experiences? No, says Clancy: "Real trauma survivors exhibit a broad range of memory impairments on this task. Recovered-memory survivors—whether the trauma is sexual abuse or alien abduction—exhibit just one impairment on this task: the tendency to create false memories."

False recall is a source-monitoring problem, an inability to remember where and when information is acquired: You think a friend told you a piece of news, for instance, but you actually heard it on the radio. "Human memory is not like a video recorder," says Clancy. "It's prone to distortion and decay over time. This does not mean that abductees are psychiatrically impaired. I don't think they should be considered weird. If anything, they're just more prone to creating false memories."

Subjects whose personality profiles indicated a high level of absorption or inclination to fantasy were the most likely to perform poorly on the word recall task. Furthermore, says McNally, every abductee in the recovered memory group described what appears to be sleep paralysis.

Clancy and McNally outlined their findings in the *Journal of Abnormal Psychology* in the fall of 2002, whittling the abduction phenomenon down to an equation of sorts. Susceptibility to creating false memories, coupled with a disturbing experience like sleep paralysis and a cultural script that allows for abduction by aliens, may lead one to falsely recall such an encounter. "You don't necessarily have to endorse these experiences to create false memories," says Clancy. "You may have just seen *The X-Files* and thought, 'That's crap,' but then you have an episode of sleep paralysis that freaks you out, and the show is still in the back of your mind."

And among people wavering about whether or not they've been abducted, hypnosis can push them to embrace this interpretation. In a 1994 experiment that simulated hypnosis, psychologist Steven Jay Lynn asked subjects to imagine that they'd seen bright lights and experienced missing time. Ninety-one percent of those who'd been primed with questions about UFOs stated that they'd interacted with aliens.

Still, if the abduction experience is a misinterpreted bout of sleep paralysis, why do abductees invest it with such emotion? A videotape of a tearful Peter Faust undergoing hypnotic regression is so powerful that Mack says he stopped showing the footage; it freaked out even nonabductees, causing many to erect "new defenses." Terror in the face of potentially false memories was one issue McNally hoped to study with abductees. This question brought him, in part, to the Divinity School conference. "I wanted to know whether people really have to be traumatized to produce a physiological reaction."

McNally collected testimony from 10 subjects with recovered memories of abduction, then confronted them with the most frightening details of their own accounts—from violent trysts to swarms of aliens around their beds. Six out of 10 subjects registered such elevated physiological reactions, including heartbeat and facial muscle tension, that they met the criteria for posttraumatic stress disorder (PTSD).

Interestingly, subjects with PTSD react physiologically only to their own traumatic experiences, but the abductee group had heightened responses to additional stressful scripts, such as the violent death of a loved one. They even reacted to positive scripts, such as viewing their newborn infant for the first time. Such reactivity, coupled with high levels of absorption, has been linked to the ability to generate vivid imagery, according to McNally. In other words, abductees are more likely to experience a traumatic—or positive—scenario as real, in part due to their fertile imaginations. They will then react to it as such. "Emotion does not prove the veracity of the interpretation," McNally concludes.

FOR MCNALLY, the most telling difference between abductees and survivors of "veritable" trauma is not physiological but attitudinal. Experiencers unanimously state that they're glad they were abducted. "There's a psychological payoff," says McNally. "This makes it very different from sexual abuse." Trauma survivors of all stripes cite positive spiritual growth, but, "no Vietnam vet says, 'Gee, I'm glad I was a POW.' "

It is understandable that memory lapses, as measured by poor performance on a lab test, pale in comparison to communication with unknown beings. And while abductees may feel assaulted by aliens, they also feel special. For that reason, "They are not trying to demystify their experience," says McNally, whose deconstruction of sleep paralysis for one woman was met with a polite smile and the exhortation that he should "think outside the box." When McNally finally broached the term "sleep paralysis" at Mack's conference, he says, "There was an awkward silence, as if someone had belched in church."

"I'm not personally interested in what Susan Clancy found," admits Bueche, for whom the memory test was "50 bucks and free Chinese food." "I don't need evidence or proof. Most experiencers are well beyond that. This is about what you can learn regardless of whether it is physically real or inter-dimensional or something grand that the mind is generating."

Mack counters that no combination of sleep paralysis and the Sci-Fi Channel explains phenomena such as alien sightings by school children in Zimbabwe who are wide-awake. "It doesn't even come close," he says. Mack's second book, *Passport to the Cosmos,* chronicles abduction as a cross-cultural phenomenon; he finds evidence of sexual and ecological parallels to American abduction reports on almost every continent.

Mack is currently at work on his third book, which examines the clash between "scientific materialism and a nonrational point of view." He increasingly distances himself from the question of whether or not aliens exist in the physical world, focusing more on a "consensus reality" that precludes us from even entertaining such a possibility. "We void the cosmos of other intelligence unless it can be proven," states Mack. On the work of McNally and Clancy in the psychology department, a stone's throw away, Mack says, "We're in different firmaments."

JOHN NOBLE WILFORD

A Tense Border's More Peaceful Past

FROM *THE NEW YORK TIMES*

The Wadi Arabah was no barrier to Lawrence of Arabia. From the village of Buseira, he and four companions rode camels down the zigzags of a steep pass through bare rock of many colors, down into the heat at the bottom of the abyss. Above, he wrote, "the cliffs and hills so drew together that hardly did the stars shine into its pitchy blackness."

Soon the travelers emerged from the narrow valley in the east and crossed miles of the open Wadi Arabah, a desolate and below-sea-level rent in the earth's surface running from the Dead Sea south to the Gulf of Aqaba. It is a section of the Great Rift Valley, which extends from southern Turkey through much of Africa.

The 110-mile-long wadi today is the tightly controlled boundary between the Negev of Israel and southern Jordan. The sparsely populated and largely undeveloped region is an object of increasing fascination among scholars who pursue what could be called the archaeology of borderlands.

To the concern of archaeologists, Israel and Jordan are casting covetous eyes on the region for a pipeline or canal to replenish the declining Dead Sea with water from the Red Sea. Unesco is considering designation of the entire Rift Valley as a World Heritage site, an action that could encourage more research in places like the Wadi Arabah.

In *Seven Pillars of Wisdom,* T. E. Lawrence described in his evocative style the depths of Arabah as "a strange place, sterile with salt, like a rough sea suddenly stilled, with all its tossing waves transformed into hard, fibrous earth, very grey under to-night's half moon."

After an arduous journey 5,000 feet down, 3,000 feet up, the travelers finally climbed out of the valley at daybreak and proceeded across a plain in Palestine to Beersheba. They covered 80 miles to join British forces. It was February 1918, in the waning months of World War I and the Allied campaign against Turkey, another time of turmoil in the Mideast.

Lawrence's adventure, crossing one of nature's more formidable obstacles to human interaction, was no mean feat, but not unheard of. History and archaeology show that for centuries, from the Stone Age through biblical times and successive empires, crossings were possible and, sometimes, regular occurrences. People traversed Arabah for trade and plunder, refuge and revenge, love and reunion.

"What we've all regarded as a border between warring nations for hundreds, if not thousands, of years was no border or barrier at all in terms of trade, resources and water," Dr. Piotr Bienkowski, an archaeologist at the University of Manchester in England, said. "Only in the last couple of generations has it become a modern fixed boundary. Only in that time have archaeologists been conditioned to think of it as a barrier."

Archaeologists may study the past, but they live in the present. For the last half-century, the boundary along Arabah has been fenced and guarded, sometimes with land mines, especially from 1948 to 1994, when Israel and Jordan were at war.

But a closer examination of historical texts and new excavations have changed minds. At a November, 2003, symposium in Atlanta, scholars acknowledged that the rugged valley, though clearly an impediment, was much less of a social and economic divide than they had thought.

The symposium, "Crossing the Rift: Resources, Routes, Settlement Patterns and Interactions in the Wadi Arabah," was seen as a modest and cautious step in an effort to expand archaeological investigation of the Arabah region and encourage more cooperation and coordination between researchers working on each side of the Israeli-Jordanian border. No one pretended that the prospects for immediate success were bright, given the volatility of Middle East politics.

The meeting was organized by Dr. Bienkowski, who excavates in Jordan, and Dr. Katharina Galor of Brown, who specializes in the archaeology of Israel. It was held with the annual conference of the American Schools of Oriental Research, an organization of scholars of Mideastern antiquity.

In the spirit of measured hopefulness, Dr. Galor opened the symposium with the greetings, "Shalom," meaning "peace" in Hebrew, and, "Ahlan wa sahlan," which in a literal translation from Arabic means, "you came to a relative of yours and you came to a place that is a wadi."

Dr. Ramadan Hussein, an Arabic instructor at Brown, explained to her that a wadi, often the site of an oasis, is meant here as a symbol of prosperity, which the greeter is offering to share with the guest.

Afterward, Dr. Galor conceded that the symposium reflected the uncertain prospects for increased collaboration in Arabah research. The meeting was originally to have been in Jerusalem, with a large delegation of Jordanian archaeologists attending. Heightened tensions forced the shift to Atlanta. One Jordanian showed up. "A completely unified enterprise does not appear realistic at this point," Dr. Galor said.

The situation reminded Dr. Thomas E. Levy, an archaeologist at the University of California at San Diego, of an Arabic expression, "Yom asal, yom basal," *one day honey, one day onions.* "The conference offered us one day of honey, and hopefully there will be many more in the near future," Dr. Levy said.

At least it was a start. Until recently, Israeli and Jordanian researchers could not attend each other's meetings and rarely shared information. Most of the researchers could sympathize with Dr. Bienkowski. "I work in Jordan and cannot explain Jordan material without looking at the whole area to explain what was happening long ago in Wadi Arabah," he said. "That has been virtually impossible."

The absence of a map of all the known excavation sites on both sides of the wadi had been a handicap, remedied in time for the meeting. A compendium of 6,000 excavation sites was compiled by computer scientists at Brown, mainly from Israeli sources.

In their papers and in interviews, the scholars elaborated on evidence that the Wadi Arabah has been traversed by people through nearly all history, Egyptians and Jews and Arabs, Edomites and Nabateans, Romans and Byzantines, incense caravans and nomadic Bedouin tribes.

Excavations at the southeastern end of the Dead Sea have found traces of roads, travelers' waste and stone remnants of way-stations of ancient trade routes leading across Arabah into the Negev Desert of present-day Israel. Dr. Yuval Yekutieli of Ben-Gurion University of the Negev said the trade routes appeared to be from the Early Bronze Age, as much as 5,000 years ago.

Archives document that Egyptians had contacts as early as the 13th century BC with a place named Edom, in what is now southern Jordan. It may be no coincidence that this appeared to be when camels were domesticated and new trade routes opened in the region.

Recent excavations by Dr. Levy and archaeologists of the Department of Antiquities in Jordan found that copper production was an important source of Edom's growth and east-west overland trade. Ancient Edom also had a major seaport near Aqaba for trade along the Red Sea.

Edom is mentioned several times in the Bible as an enemy of the Israelites. Scholars now think Edom was not a centralized state, but more of a confederation of tribes. In any case, the Edomites did not stay put on their side of the wadi. Their god, Kos, was sometimes worshiped west of the wadi, indicating that some Edomite tribes had settled there.

By the late fourth century BC, people known as the Nabateans were living east of Arabah. Merchants in the incense and spice trade, the Nabateans plied their commerce on routes crossing Arabah to Gaza and others taking them to Damascus. A beautiful relic of Nabatean incense wealth is the first century BC architecture of Petra, their seat east of Arabah. Nabatean influence was wide. Their pottery and other artifacts have been found at many sites in Israel. A Jewish king, son of Herod the Great, married a daughter of the Nabatean king, perhaps to cement economic ties across the border.

Dr. Clinton Bailey, a scholar at Hebrew University, has lived with Bedouin tribes in the region and translated their poetry. His research, he said, shows that Arabah has at times been a barrier and an interface. Tribes of the nomadic pastoralists worked both sides of the wadi. But for a long time, Dr. Bailey noted, people west of the wadi "saw danger coming from the east, saw the Arabah as a route of conquest and raids."

Among the most significant finds are copper mines and processing factories. Some have been uncovered on the western side, notably at Timna. The most extensive remains of a copper industry are along the eastern edge, in Jordan. Thirty miles south of the Dead Sea, in its Faynan area, a team led by Dr. Levy excavated a 5,000-year-old copper-processing center that experts said was one of the largest copper sources in the eastern Mediterranean then. In the rubble at Khirbat Hamra Ifdan, excavators uncovered stone hammers, anvils and crucibles, as well as molds for copper axes and chisels.

Another discovery by Dr. Levy's group—including Dr. Russell B. Adams of McMaster University in Ontario, Dr. Andreas Hauptmann of the German Mining Museum and Dr. Mohammad Najjar of Jordan—is a copper factory from the biblical era.

The site at Khirbat en-Nahas, east of the wadi, and evidence of trade routes leading from it, Dr. Levy said, showed "conclusively that the Wadi Arabah was never a barrier to social and cultural interaction." An analysis of the distinctive "fingerprints" in the metal showed that copper from east of the wadi was traded widely in Egypt and Israel.

Even with the discoveries, Dr. Adams cautioned that many problems remained in "understanding the wadi in its entirety and in terms of human interactions across it."

But like Lawrence of Arabia, Dr. Levy has demonstrated again that the Wadi Arabah is no uncrossable barrier. In 1997, he led a group of Jordanian, Israeli, American and German researchers on a trek across the wadi, following an ancient trade route from the copper center at Faynan to an exit leading to Beersheba, in Israel.

The expedition used two modes of transportation, donkeys and the trekkers' own feet. The donkeys carried saddlebags loaded with copper to re-enact ancient treks. The journey across took only six hours. It was enough, Dr. Levy said, for researchers from different and sometimes hostile countries to develop collegial ties.

Describing the trek at the symposium, Dr. Levy said: "If anything, the relatively short distances in the Arabah and the relative abundance of springs within a day's walk of each other ensured that trade, exchange and social interaction was always possible whether conducted by foot, donkey or camel. The chief impediment would have been social relations between the people living in the region—just like today."

Tom Bissell

A Comet's Tale: On the Science of Apocalypse

FROM *HARPER'S*

> I don't really think that the end can be assessed as of itself, as being the end, because what does the end feel like? It's like trying to extrapolate the end of the universe. If the universe is indeed infinite, then what does that mean? How far is all the way and then if it stops what's stopping it and what's behind what's stopping it? So "What is the end?" is my question to you.
> —DAVID ST. HUBBINS,
> *This Is Spinal Tap*

Apocalypse

WE HAVE ALL READ IT, or think we have. It does not take long to read: forty-five minutes, by my reckoning. The Bible's concluding chapter begins directly, as a letter, quickly sashays into garish incoherence, occasionally breaks into startling verse, and above all promises to show "what must soon take place." It has been used most often, by religious charismatics, as a kind of divine itinerary: Jesus catching the 7:15 out of he ven, arriving in flames at Jerusalem's holy Mount anon. Although America enjoys an early tradition of apocalyptic thinkers (the saintly Roger Williams among them), a visiting Anglican priest named John Nelson Darby was the first to smuggle into the New Canaan an organized vision—known as Premillennial Dispensationalism—centered around the book's theology of blessed annihilation. That was in 1862. An American apocalypse, perhaps not entirely coincidentally, was in the process of erupting. Despite chronic humiliations, Premillennial Dispensationalists are with us still. A recent poll disclosed that 59 percent of the American public believes that the events projected by one John of Patmos will come true. Ancient Christians apparently shared as strong a belief in the sturdiness of John's prophecies. An annoyed Augustine of Hippo advised that "those who make calculations" based on John's auguries should "relax your fingers and give them a rest."

The vernacular name by which this queer book is often known, Revelations, is wrong. It is Revelation, or The Revelation to John (and note that preposition). The book's Greek title is *Apokalypsis,* or Apocalypse, which at its peaceable etymological nativity meant simply "unveiling." Revelation comes to us from the Vulgate Bible, the translators of which bowed *Apokalypsis* into *Revelatio,* the nearest Latin equivalent. Drawing partly on the "small" apocalypse in the book of Daniel, partly on Jesus' premillennial dispensationalist discourse on the Mount of Olives, and partly on God knows what, Apocalypse has arguably influenced the Christian worldview more than any other book in the Bible—surprising for a gloomy phantasmagoria that the Good Book's fourth-century compilers very nearly omitted. Yet here we are, still helplessly attempting to clear away its two millennia of murk. The same bewildering poll that unveiled a 59 percent rate of American acceptance of Apocalypse also revealed that a quarter of Americans believe that their Bible—despite a conspicuous textual absence of airplanes, skyscrapers, or Muslims—predicted the horrific but hardly apocalyptic attacks of September 11.

In Apocalypse, most biblical scholars now agree, John of Patmos is actually describing a situation that, through archaeology and textual analysis, we are more or less able to piece together. Some churches are "falling asleep," others are behaving licentiously, and a great beast (probably Nero Caesar, whose name adds up to 666 when Hebrew letters are given numerical values; this is called gematria) doth loom astride the scattered Christians' beleaguered world. Apocalypse is an angry, fatherly chastisement that combines the "cosmic battle" scenario familiar throughout the ancient world with a more newly minted Christian persecution narrative. In other words, Apocalypse makes us privy not to the future but to John of Patmos's recent past and immediate present—a trick of perspective, often used by biblical authors, called *vaticinia ex eventu,* history disguised as prophecy. Apocalypse's numerological *idée fixe*—the Alpha and Omega, the slain Lamb's seven horns, seven seals, seven bowls, four horsemen, 666 itself—is symptomatic of nothing more than the ancient world's obsession with "whole" and "incomplete" numbers. Rest easy, Christians. Relax your fingers.

LEST WE REST TOO EASY, however, here is a numerical sequence that does have bearing on the end of the world: 1950 DA. This is the designation given to an asteroid that after fifteen more near-Earth passes may eventually collide with our planet. If this rock fulfills its most dreadful cosmic destiny, John's prediction of a sky vanishing like a scroll rolling up into itself, his stars falling to Earth, his black sun and earthquakes and lakes of fire and bloodred moon, will all come quite horribly to pass, with one exception: there will be very few Christians, very few people, left to exult in this much-delayed fulfillment of prophecy.

Bang, Whimper, Fire, Ice

NOT EXCEPTING ROBERT FROST'S famous eschatological either-or, the world will end in fire *and* ice. Our sun, now enjoying the solar equivalent of middle age, will, somewhere between two and five billion years from now, run out of fuel and "go nova." It will first become what is known as a red giant, a faltering star with a surface temperature much lower than that which it currently has. This process will unfortunately cause the sun to expand many times beyond its present size, incinerating everything between it and Jupiter. The sun's burning wave of helium will be traveling tens of thousands of miles per second, and on Earth the sky will pass from blue to yellow to the fire of atmospheric ignition in roughly the time it takes to blink. Once this terminal solar spasm has run its course, the sun will collapse into a small, cool, extremely dense star known as a white dwarf. The remains of the inner solar system's flash-fried orbital matter will then be encased within icy spherical coffins hundreds of miles thick. Since very few species have ever managed to achieve a life span anywhere near even one billion years, it is extremely unlikely that recognizable inheritors of humanity's mantle will be witness to this celestial endgame. "Going nova" is, for now, an intellectual indulgence for astrophysicists and neurotic civilians whose pessimistic affirmations are most comfortably expressed in cosmic terms.

What, then, do we mean by The End? If we mean The End of the World Itself, we need not concern ourselves. Setting aside the aftereffects of our sun's demise fifty million centuries hence, Spaceship Earth is stubbornly resilient. In geologic time, it is probably indestructible. Do we mean The End of Life? If so, of what kind of life? Three and a half billion years ago the primordially molten Earth cooled enough to allow for the stabilization of its chief chemical components. The result was swarms of prokaryotic cells that held planetary sway for two billion years. One and a half billion years ago these simpletons became eukaryotic cells, replete with fancy new cytoplasm and nuclei. Another billion years passed before the occurrence of anything resembling multicellular development. The story of life on Earth is largely the story of stromatolites, organisms not likely to have believed that the god of their understanding reserved a special place for each of them in heaven. Do we mean by The End of the World, then, The End of Multicellular Life? If so, the world has, by any practical definition, already ended. Two hundred and fifty million years ago, during the later Permian period, 95 percent of all Earth's life was suddenly eradicated. Do we mean something as anthropocentric as The End of Human Life? If so, how many of us have to die to signal the collapse of the human endeavor? Perhaps we should ask the ghosts of European Jewry. For the citizens of Hiroshima and Nagasaki, the world certainly appeared to end on August 6 and August 9, 1945. For the people of China the world ended with equally inarguable force in 1556, when the worst earthquake in recorded history tore open, lifted up, and smashed back into place the Shensi province, in the process snuffing out 800,000 lives. More dramatically, 73,500 years ago what is known as a volcanic super-eruption occurred in modern-day Sumatra. The resultant volcanic winter blocked photosynthesis for years and very nearly wiped out the human race. DNA studies have suggested that the sum total of human characteristics can be traced back to a few thousand survivors of this catastrophe.

Earth itself is the most ambitious mass murderer in the galaxy. Its pleasant, cloudy face should be slapped onto a WANTED poster, and soon: this lunatic is out to get us. As the volcanologist Bill McGuire notes in *A Guide to the End of the World,* an elegant volume of nervous-breakdown-inducing persuasiveness, "The Earth has been around just about long enough to ensure that anything nature can conjure up it already has." As recently as 1902, Mount Pelée in Martinique exploded, vaporizing the town of St. Pierre with a glowing avalanche of skin-melting gas and lava flowing as fast as a cheetah at full sprint. Thirty thousand people died; the blast's two survivors were likely to have regarded any notion of the world's end with seared, unblinking eyes. Earth is cratered with somewhere between 1,500 and 3,000 volcanoes, and at least one explodes each day. Most of North America's volcanoes have been dormant for centuries—one particularly angry specimen detonated 450 million years ago and evaporated an area the size of Egypt—and this is certainly good news. The bad news is that the volcanoes that have gone speechless for the longest periods of time tend to have the most to say when they finally break their silence. The long-dormant Indonesian volcano Tambora blew off its peak in 1815, with an explosion heard 900 miles away. Twelve thousand Indonesians died as a direct consequence of the eruption, and another 80,000 perished due to famine. The atmosphere was crowded with so much ash and sulfur that 1816 became known as the "year without a summer." England saw its heaths frosted through July, allowing Mary Shelley the proper frame of mind to complete the wintry *Frankenstein*.

THAT SUCH NATURAL DISASTERS SEEM, in the Western mind, exclusive to primitive people in far-off lands is mostly a fluke of our current distributional good fortune. This quietly unapologetic stance, which braids laissez-faire eugenics with Western exceptionalism, is in fact a highly callous form of disaster denial, for yesterday's village-erasing lava flow in Sumatra is tomorrow's super-eruption in British Columbia. Human civilization as a whole has grown amid a long breather from the kinds of natural calamities that, 10,000 years ago, left our freshly de–Ice Aged planet broiling beneath junglish centigrade from Ellesmere Island to Admundsen-Scott Station. It would probably require a psychiatrist magically privy to the workings of the mass mind to explain why we as a species seem resistant to believing that such disasters could recur, especially when many feel so ominously near. The planetary conditions we are today seeing unfold—alarming rises in human population, sea level, and temperature (in 100 years Earth will be hotter than it has been at any time in the last 1,500 centuries)—could very well be the overture to nature's awful new symphony.

Yes, the old doomsday warhorses of overpopulation and global warming (which actually should be called anthropogenic warming) are still bedeviling us. Little more can be said of global warming, other than to note that it is well under way and that those who maintain otherwise are the meteorological equivalent of creationists. In the case of overpopulation, no one can deny that the world's inhabitants doubled between 1960 and 2000. Whether this is merely bad or catastrophically bad is still unclear. Equally unclear is whether overpopulation (a term too often used as code for too many brown and yellow people) will result in a long-feared drain of world resources. But the truth remains that this sudden explosion of population is, most basically, unprecedented in human history.

One runs the risk of sounding like an idiot, or worse, when pondering the sandwich-board finalities of The End. Consider the most famous Cassandra of the 1970s, Hal Lindsey, author of *The Late Great Planet Earth,* the best-selling book of that deeply regrettable decade. (It still, somehow, sells around 10,000 copies a year.) His predictions of the rise of a single world religion, a Soviet-Ethiopian invasion of Israel, and the obliteration of Tokyo, London, and New York, all systematically unfulfilled, have made him a laughingstock along the lines of pet rocks. It is with a sense of near torment, then, that I report Lindsey's Reagan-era sinecure as a consultant on Middle Eastern affairs. Our current decade's disseminators of global doom, Jerry B. Jenkins and Tim LaHaye, authors of the projected twelve-novel cycle *Left Behind,* are hardly improvements. The series' ninth installment, the accurately titled *Desecration,* was the best-selling novel of 2001. These jaw-droppingly substandard books peddle a religion of fear and despair (passenger-laden airliners, whose faithly pilots are Raptured up to heaven, are allowed to crash), partake of some truly shocking Jew-baiting (Nicolae Carpathia, the books' antichrist, lately of the United Nations, is helped along by a sinister cabal of "bankers"), and pile up body counts with slasher-flick glibness (the nuclear immolation of Chicago merits a single paragraph). And just as Hal Lindsey filled the mental shoals of Ronald Reagan with his insights into the Middle East ("If you stumble over some of these unpronounceable ancient names, glance at your daily newspaper and notice the tongue twisters assigned to people today"), Tim LaHaye served as the co-chairman of Jack Kemp's 1988 presidential campaign.

Apocalypticism, it hardly need be said, drags out of humanity all that is small and terrible and mean. It also drags out what is worst about God, for whom love seems an infrequent mood. But the apocalyptic vision speaks to us, and has spoken to us, for thousands of years. It has allowed the quatrain-scribbling Nostradamus to live indefinitely off the obscurantist residuals of human anxiety. It has allowed some in the Muslim world to regard figures as disparate as Pope Innocent III and George W. Bush as the long-feared Dajjal. It has allowed George W. Bush, arguably the worst president in American history, a surreally nonexistent pretext for world war. It seems we are all going a little nova. In July of 2002, 13,500 men and women of the army, navy, air force, and marines engaged in a war game called Millennium Challenge. That exercise, the largest of its type ever conducted, was based on a rogue military commander staging a coup in an earthquake-riven Middle Eastern nation. Make that millennium-challenged: the rogue military commander won, until the war game's rules were reset.

Is a war game rigged to ensure victory a reasonable depiction of The End? Or does The End consist merely in extrapolating the size of the volcano needed to turn Jakarta into a cinder or calculating how many feet of water New En-

gland will be sitting under once the polar ice caps melt? Is it discerning the arctic climate of Great Britain once the Gulf Stream vanishes or projecting the
megadeath of a nuclear exchange between Islamabad and Delhi? It will be no
easy thing to eradicate six billion human beings, the most wide-ranging, adapt-
able, and notoriously intelligent creatures to which the planet has ever played
host. But whereas human beings are tough, the civilizations they form are
appallingly delicate. Here is the buried psychological certainty—*I may die, but
human culture will endure*—that the apocalyptic vision upends.

THUS WE HAVE ENTERTAINED, and still entertain, The End at the hand
of God. We have been plagued, and will be plagued, by The End at the hand of
the planet itself. What seems most likely is that The End of the World will mean
the reduction of humanity to a dangerously puny number, whereupon our
planet will become another planet altogether, something we will live upon but
that will no longer belong to us. Perhaps that future will be simpler and more
peaceful, perhaps toothier and rougher. *Star Trek* or *Mad Max*? If we do proceed apace and engage in a global swap of ICBMs, who among us could argue
that we did not have it coming? But imagine the different, far less personal end
embodied by objects such as 1950 DA. Who among us would argue that we had
that coming?

This new End means the loss of God, and not only because it creates a kind
of philosophical yard sale in which everything must go. No, God will be discarded long before The End truly begins. We have already rummaged off not a
few versions of Him. The hirsute old thunderbolt-hurlers to whom we long
paid tribute—priapic Zeus, testy Yahweh—cannot map the genome, or tell us
our past, or even explain our future. Only we can do that. "Playing God" previously meant the ability to take life, a feat we too can now achieve with spectacularly divine wrath. The old God, whatever his alias, is dead, and the God we
are left with today is of little discernible help. Or perhaps God is nothing more
than a mile-wide chunk of cosmos-wandering silicate, serenely floating right
toward us. Was it all a mistake?

I Heard the Learn'd Astronomer

NASA'S DEEP SPACE NETWORK antennae at the Goldstone Deep
Space Communications Complex are spread out along twenty-six miles on the
grounds of Fort Irwin, a United States Army training center in the desert
wastes between Los Angeles and Las Vegas. I am traveling to the complex,

which is fenced off from the surrounding military base, with Steve Ostro, a radar astronomer who works out of NASA's Jet Propulsion Laboratory (JPL), in Pasadena, California. Ostro is a handsome, southern-California-fit man in his early fifties. His resemblance to a more dashing version of Russell Johnson's Professor from *Gilligan's Island* is spoiled very slightly by his glasses, which although not unflattering are as thick as bulletproof glass. As we chat we pass by small road signs that say things such as TANK XING, billboards that read THE ABCS OF SAFETY: aLWAYS bE cAREFUL, and hard, dried-up lakebeds that serve as landing strips. Often, Ostro tells me, he has seen military exercises, some of them live-fire, playing along these bleak horizons: Black Hawk helicopters cruising 100 feet off the ground at 130 miles per hour, tank columns advancing upon coyotes and gophers while huge volleys of artillery boom across the vacant desert sky.

Goldstone's centerpiece antenna, the DSS-14, was built in 1966 to receive communications from, and plot the movements of, early NASA missions to Mars. The monolithic dish dominates the surrounding, studies-in-brown terrain due to both its twenty-four-story height and (despite a crow infestation) its gleaming polar whiteness. When pointed straight up, as it is now, it strongly resembles a chandelier. The nine-million-pound antenna rests atop a round pedestal and has a pointing precision measured in something called millidegrees. Small, anonymous buildings wreathe the antenna, some housing dozens of tall metal lockers that contain telemetry equipment used to communicate with deep-space vehicles, others crammed with chugging generators that provide the antenna enough wattage to power a small town. Inside these buildings it is kept very cold—a little less than fifty degrees—to prevent the computers from overheating. Not surprisingly, a low and vaguely doomed hum pervades Goldstone's every nook and turn, rather like what one hears aboard a cruising 747, only much louder, and many who work here complain of headaches. After two days I will be complaining of one myself.

For the last several years Ostro has been coming to Goldstone to map by radar our solar system's larger known asteroids. The DSS-14 sends out a beam far too thin to discover new asteroids. That duty is handled by the Near-Earth Asteroid Tracking project at JPL, which reports its findings to the Minor Planet Center, in Cambridge, Massachusetts. If the observed object is a new asteroid—occasionally older asteroids go missing and are mistaken on reappearance for new ones—MPC gives the object a number based on the year and month and order in which it was discovered. (1950 DA, for instance, was discovered in 1950 during the second half—each half-month is given a letter—of

February. The A indicates that it was the first asteroid found in that half-month.) NEAT observes the new asteroid until its orbit can be reasonably determined, at which point Ostro takes over. Powered by a half-million watts, the antenna sends out into space a beam with the angular resolution of the human eye. The beam hits the targeted asteroid uniformly, scattering in every direction; only a small portion of the beam's energy ever makes it back to Goldstone. But this tiny fraction provides Ostro and his colleagues with enough information to determine the asteroid's velocity, size, and likely structural components. Often enough, they are able to use the collected data to create a grainy but nevertheless beautifully revealing image of the object.

As we walk into the lab housed in the antenna's pedestal to meet with Ostro's colleagues, he tells me he finds radar astronomy "an unbeatable experience." What before had been only an infinitesimal point of light attached to an ungainly number suddenly becomes a tiny, detailed world. I begin to understand his remark, made way back on Interstate 15, that he stopped reading the sci-fi novels he loved as a teenager when the science he was involved in became more interesting to him than fantasy.

Inside the pedestal I meet Lance Benner, John Giorgini, and Ray Jurgens, JPL scientists highly adept in the different areas of astronomy and planetary science that allow the team to apply an astonishing interpretive breadth to the antenna's radar readings. Jurgens, the oldest of the men, seems to be providing a good deal of support merely by standing on the lab's fringes and quietly observing. All of them are distracted, as the antenna's transmit-receive cycle is about to begin. I stand to the side, staring at a few taped-up computer-generated images of Goldstone-mapped asteroids.

Very little is known about asteroids. The spin states, shapes, geological compositions, surface characteristics, and collisional histories are, for the vast majority of identified objects, still a mystery. Only with radar imaging is much of this data, accomplished one object at a time, finally becoming clear. The asteroids' irregular shapes give each something resembling a personality: Kleopatra is shaped like a dog bone, Eros like a rutabaga, Geographos like a turd. Some rotate evenly, others like badly thrown footballs.

The lab's equipment appears, to me, strangely antiquated. Somehow our technology improves but gets no closer to the touch-screen sleekness of cinematic futurism. Little red and white lights blink on the hulking computers' faces alongside small screens active with greenish waveforms. A spray of cables hangs from seemingly every panel. Jurgens, who has worked at Goldstone since the 1970s, walks over and explains that much of this equipment is twenty years old. Later inquiries as to how adequately NASA funds Goldstone's radar astronomy work—it is about one ten-thousandth of NASA's overall budget—will result in meaningful silences.

Ostro escorts me to another computer at the room's far end, the real-time sawtooth display of which will soon show us the electromagnetic Doppler frequency the antenna is receiving back from the asteroid. This information will be used, Ostro explains, to calculate the asteroid's orbit uncertainty. "We go through this process where we have a projected orbit. We see how good the prediction was, and then we make a better orbit and a new prediction of uncertainty. There's always uncertainty, and that's one of the really interesting domains of this whole problem. What is the uncertainty, and how do we reduce it? That's why we're here."

I am here, I remind him, to find out about the chances of one of these asteroids colliding with Earth. But this is where the uncertainty comes in, he tells me. Every asteroid travels along a path we can determine using orbital trajectories, but within that trajectory there exists an error ellipse in which we cannot be sure where the asteroid will be. This ellipse can be many hundreds of thousands of miles in width, which makes reducing the uncertainty of an object that passes near Earth that much more important. Asteroid 1997 XF11, the asteroid we will be observing today, has a curious history of uncertainty. Initially, 1997 XF11 was predicted to have a small but nonzero chance of hitting Earth in 2028. Whether this was due to hasty "back of the envelope" calculations or a real uncertainty in its error ellipse is still debated. What is known is that the media frenzy was so immediate ("Killer Asteroids!") that NASA created the Near-Earth Objects Office to handle future impact threats. It was later determined that 1997 XF11 had no chance of hitting Earth, and before the day ends we will know everything else there is to know about this defanged rock.

"A lot of the confusion about this topic," Ostro goes on, "ultimately comes down to miscommunication. All of this is unfamiliar and intrinsically arcane and inaccessible and beyond the experience of humanity." (A sample from a paper Ostro gave me: "This is simply saying that a survey system with a limiting magnitude $m_{lim} = 20$ will achieve the same completeness of absolute magnitude $H = 20$ objects as a system with $m_{lim} = 19$ will achieve of $H = 19$ objects.") I ask Ostro if he personally worries about the day they discover an asteroid that has a high probability of hitting Earth. He is silent for such a long time that I ask again. "Let me rephrase your question," he says. "Is there a God?" After some uncomfortable laughter on my part, Ostro tells me about 1950 DA, the only large asteroid currently known that has a nonzero chance of colliding with Earth before the next millennium. If it does strike, it will impact the North Atlantic just off the U.S. coast in March of 2880. "It's a little bit of a stretch to say it might hit the Earth—the probability is 1 in 300—but it's the most dangerous object we know. Now, do we care about that? Should anybody care at all about the fact that an asteroid might hit Earth nearly a millennium from now?"

I imagine standing in a room with 300 people, then being told that one of us will be taken outside and shot. I tell Ostro that I think I can care about that.

"What if I had said, 'Well, we found an object that has a pretty good chance of hitting the Earth in 500,000 years.' Would that concern you? Should we care about that?"

I admit that I have a hard time gathering the emotional momentum that allows my concern to travel ahead a half million years.

Ostro nods. "What it comes down to is that this is a very new kind of topic, and it's hard to get one's bearings thinking about it, much less for society to decide whether to worry and spend money on it. And if so, how much, and how?"

The lab's small, encaged red light begins to flash in alert: the antenna is finally transmitting its half-million-watt, pencil-thin beam of energy seven million miles into deep space. Its round-trip time back to Goldstone will take a little under eighty seconds. As we wait, I find myself thinking of Whitman's "When I Heard the Learn'd Astronomer." In the poem Walt grows so "tired and sick" of an astronomer's "charts and diagrams" that he goes out "In the mystical moist night-air, and from time to time,/Look'd up in perfect silence at the stars." I wonder what poem he might have written had he known that some of those stars had the potential to end poetry, and everything else, for all time.

HOW DID WE GET HERE? In 1178, a monk in Canterbury, England, recorded the testimony of two men who witnessed a "flaming torch" spring up off the face of the moon, which "writhed, as it were, in anxiety," then "took on a blackish appearance." What these men saw, some scientists believe (the issue is debated), was the formation by an asteroid collision of the moon's youngest known crater, Giordano Bruno, named in honor of a defrocked Italian philosopher-priest. The explosion had the estimated force of 120,000 megatons, equal to 120 billion tons of TNT. Hiroshima was a mere 15 *kilo*tons. The greatest man-made explosion in history, a Soviet nuclear test on the Arctic island Novaya Zemlya in 1962, was 60 megatons. If every nuclear device on the planet were somehow to explode at the same moment, no more than 20,000 megatons would be unleashed. The formation of Giordano Bruno, if that is indeed what the two witnesses saw, marked perhaps the first time in recorded history that human beings observed what is now known as a large-body impact. The next would occur more than 800 years later, when two dozen fragments of a shattered comet would explode on Jupiter. Not even the intervening centuries of scientific advancement would allow us any true comprehension of the destructive potential of large-body impacts. Faced with the effects of 20 megatons of explosive energy for every man, woman, and child on Earth, the mind is quickly beaten into something misshapen and medieval.

The term "asteroid" means "like stars," stars being what earlier humankind most often mistook asteroids for. When an asteroid breaches Earth's atmosphere it becomes a meteor; when it strikes Earth's surface it is called a meteorite. A "shooting star" is typically envisioned as a midsized burning chunk of speeding rock, when in fact most shooting stars are no bigger than a grain of sand. The speed at which they travel, and the opposing force of Earth's atmosphere, cause the particles to explode. The streaks of light we see in the night sky are these particles' violently released energy.

The solar system's primary asteroid repository whirls in a formation known as the asteroid belt, which lies between Mars and Jupiter, the latter possessing our system's second most powerful gravitational force, after the sun. Most asteroids are the fragmentary remains of the same cosmic *deus ex nihilo* that discharged rock and matter across the galaxy several billion years ago. In our solar system alone, as many as a trillion pieces of debris once floated around the perpetually contained hydrogen explosion we know as the sun, most of which were bashed into oblivion by collisions. Only nine of the survivors have been dignified with the name of planet, and only one is known to have developed complex forms of life. This did not make them invulnerable. A Mars-sized object slammed into the nascent Earth billions of years ago, for instance, and the drama of its effect can be appreciated by the fact that it threw off a huge, wounded, molten glob that froze, was captured by Earth's gravity, and eventually became the moon. Because of this ancient demolition derby, the solar system is presently a much more open place, and collisions are far less frequent.

The first identified asteroid, Ceres, was found in 1801. By 1900 astronomers had located 462 more. All but one (Eros, discovered in 1898) were asteroid-belt objects. In the last 100 years, 150,000 more have been pinpointed and the orbits of 52,000 accurately determined. Most of this orbital surveying was accomplished after 1968, when the asteroid Icarus passed within four million miles of Earth and first caused astronomers to ponder the possibility, if not the likelihood, of large-body impacts.

Asteroids are categorized into several classes. S-types, which dominate the inner half of the asteroid belt, are composed of stone and silicate materials. C-type asteroids, which take up most of the outer half, are dark rocks rich with complex organic compounds called carbonaceous chondrites. P- and D-type asteroids are the farthest away and, consequently, the most compositionally mysterious. The belt's nearest asteroids are M-types, highly reflective iron-nickel objects that scientists believe are tantalizing atavists of planetary cores. In Ostro's office at the Jet Propulsion Laboratory he handed me a black-and-gold chunk of a billion-year-old M-type asteroid. Although it was only a little larger than a compact disc, my hand nearly hit the floor. Its density, Ostro explained, was twice that of any object of terrestrial origin, and to cut it open would reveal an interior as bright as freshly forged steel. The smallest known M-type asteroid, 3554 Amun, with a radius of only 500 meters, is thought to contain $1 trillion worth of nickel, $800 billion worth of iron, and $700 billion worth of platinum. Since towing it back to Earth would be counterproductively expensive, and likely to result in the collapse of the world market for fine metals, these figures are, for now, lit with little more than the neon of sci-fi dreams.

In its journey around the sun Earth passes through the orbits of twenty million asteroids. Many of these Earth-crossers are called near-Earth asteroids. NEAs much smaller than 100 meters wide are basically undetectable but for a fluke of stargazing luck; unfortunately, an object of only, say, 90 meters possesses the collisional capability of roughly 30 megatons of explosive energy, a figure that is dreadful but globally manageable. NEAs larger than 100 meters are thought to number 100,000, a fraction of which have been located; in the event of an impact these could effect serious global climate change. Around 20,000 NEAs are large enough, individually, to annul a country the size of the Czech Republic. The number of NEAs bigger than one kilometer in diameter is currently thought to be around 1,000. At astronomers' current rate of detection—roughly one a day—a survey of the entire population of one-kilometer NEAs will be complete within the next decade.

This one-kilometer threshold is important, for asteroids above it are known as "civilization-enders." They would do so first by the kinetic energy of their impact, striking with a velocity hitherto unknown in human history. The typical civilization-ender would be traveling roughly 20 kilometers a second, or 45,000 miles per hour—for visualization's sake, this is more than fifty times faster than your average bullet—producing an impact fireball several miles wide that, very briefly, would be as hot as the surface of the sun. If the asteroid hit land, a haze of dust and asteroidal sulfates would enshroud the entire stratosphere. This, combined with the soot from the worldwide forest fire the impact's thermal radiation would more or less instantaneously trigger, would plunge Earth into a cosmic winter lasting anywhere from three months to six years. Global agriculture would be terminated, and horrific greenhousing of the climate and mass starvation would quickly ensue, to say nothing of the likely event of world war—over the best caves, say. In the event of a 10-kilometer impact, everything within the ocean's photic zone, including food-chain-vital phytoplankton, would die, but this would hardly matter, as the deadly atmospheric production of nitrogen oxides, which would fall as acid rain, would for the next decade poison every viable body of water on Earth. Chances are, however, that the impact would be a water strike, as 72 percent of meteorite landings are thought to have been. This scenario is little better. A one-kilometer impact would, in seconds, evaporate as many as 700 cubic kilometers of water, shooting a tower of steam several miles high and thousands of degrees hot into the atmosphere, once again blotting out incoming solar radiation and triggering cosmic winter. The meteorite itself would most likely plunge straight to the ocean floor, opening up a crater five kilometers deep, its blast wave cracking open Earth's crust to uncertain seismic effect. The resultant tsunami, radiating outward in every direction from the point of impact, would begin as a wall of water as high as the ocean is deep. If a coastal dweller were to look up and see this wave coming he or she would be killed seconds later, as it would be traveling as fast as a 747. Of course, these are all projections based in physics, and can be scaled either slightly up or slightly down in their potential for global destruction. As the paleontologist David M. Raup puts it, "The bottom line is that collision with a . . . one-kilometer body would be most unpleasant."

Although one-kilometer impacts (at least several thousand megatons) are thought to occur once every 800,000 years, with 200-meter objects (1,000 megatons) striking once every 100,000 years and 40-meter objects (10 megatons) striking once every 1,000 years, only a handful of professional and amateur astronomers are currently watching the skies. Nearly half of the asteroids believed capable of destroying one quarter of humanity remain uninventoried. Not until 1998 did the U.S. Congress direct NASA to identify, by 2008, 90 percent of all asteroids and comets greater than one kilometer in diameter with orbits approaching Earth. Unfortunately, the government agency—of any government, anywhere—that would react to and be expected to deal with the likelihood of an asteroid impact does not currently exist. The impact threat is what Ostro calls "low probability and high consequence," and bureaucracies scatter like roaches from the kitchen-bright possibility of severe consequences. We need only to consider the disgraceful games of administrative duck-duck-goose played in the aftermath of comparatively smaller disasters, such as the terrorist attacks of September 2001, to recognize the federal unwillingness to counter its own congenital laxity.

Nonetheless, as I wait with Ostro, Benner, Giorgini, and Jurgens for 1997 XF11's first measurements to appear on-screen, I experience something like patriotism. The United States is currently the only nation in the world doing anything about the possibility of asteroid impacts. I am standing with a group of interstellar Paul Reveres. When I mention this to Ostro he shrugs. "The world owes a great deal more to the United States than is commonly supposed," he says, staring at the screen. "And the United States should be very proud of itself for supporting the research it does, and being the first to take this seriously."

The bandwidth begins to come alive, and after a little while Ostro is excitedly pointing things out. "It looks like it's a fast rotator; it looks like it's more or less spheroidal; it's not an elongated object. We guessed its size at a little more than a kilometer, and that looks to be about right." While Benner begins the slow process of creating a pixelated image of 1997 XF11, Giorgini sits down at a computer console to exploit the new radar information to reduce the uncertainties of the asteroid's orbit by a factor of 5,000, using its current position to integrate its orbit backward and forward in time. Jurgens laughs and says that twenty years ago this would have required two years of computation. When I ask if that resulted in more or less error, Jurgens says, dryly, "There was a lot of error."

Giorgini's computer is essentially putting to use every practical thing that human beings have learned about mathematics and physics over the last 1,000 years. While we wait for it to provide a table of the asteroid's journey through time and space, I ask Giorgini if he ever worries about impacts. "It's unlike almost all other natural disasters," he says. "We can't do much about a hurricane, and we can't do much about volcanoes, but there's a predictability to asteroid impacts that will give us an interval, and the interval is comparable to a human lifetime. Some guy working alone in his basement could design some killer bacteria without anybody knowing about it. Whether an asteroid hits us before then, I don't know. You can't worry about everything." When I ask about the day 1950 DA's nonzero likelihood of impact came up on Goldstone's screens, he gives his head one brisk shake. "That was very exciting."

Suddenly Ostro tells me that if 1997 XF11's impact hazard unexpectedly comes up nonzero I will be escorted into the desert and left for the coyotes. I ask whether impact "cover-ups" fall under the heading of right-wing or left-wing conspiracy. "Both wings flap together," Ostro says.

The information we have been waiting for begins to unscroll in several columns down Giorgini's screen. "So here we have the close-approach table," he says. I determine, privately, to call it the Holy Shit Table. "Notice these are all zeroes from the year 1900 up through the year 2100. And here in 2028"—its former impact year—"you can see it's about two and a half lunar distances from Earth." He executes a few quick keystrokes and brings up another table showing the asteroid's close approaches, or "planetary encounters," every year from 1627, a year before Salem was founded on Massachusetts Bay, to 2228. Nothing but zeroes in the Holy Shit Table, not only for Earth but for the moon, Venus, and Mars as well. Until then, at least, we are safe from 1997 XF11. This, Ostro says happily, is as close to time travel as we are likely to get. Thanks to one radar reading we have been awarded the virtual travel diary and itinerary of an object seven million miles away. We are safe.

The Giggle Factor

IMPACT THEORIES are not new. One of the first scientists to argue in favor of them was Dr. Grove Karl Gilbert, in 1893, though he went only as far as placing past asteroid impacts on the moon. The resistance to dealing with the implications of Earth-based impacts, however, is almost as old as science itself. The planetary scientist John S. Lewis, author of *Rain of Iron and Ice* (somewhat plaintively subtitled *The Very Real Threat of Comet and Asteroid Bombardment)*, credits this resistance to the "giggle factor," a "half-suppressed hysteria that arises from an emotional inability to deal with the truth." Whereas Isaac Newton was obsessed with the end of the world, dusting relentlessly for the sulfuric cultural fingerprint of the antichrist, he dismissed the (even then) growing evidence for large-body impacts, believing that God had put everything in its proper celestial place. When a meteorite struck Weston, Connecticut, in 1807, two Yale professors verified the strike as extraterrestrial. Thomas Jefferson, after parsing their report, allegedly said that he "would find it easier to believe that two Yankee professors would lie, than that stones should fall from the sky."

Earth is home to more than 170 known impact craters, some as old as two billion years. Erosion has undoubtedly erased dozens more. For many years these craters, including the startlingly well preserved Meteor Crater in Arizona, were said to have resulted from "crypto-volcanic" (hidden volcanic) activity, though some argued that this was impossible. On account of the controversy surrounding Meteor Crater's origins, maps made of Arizona prior to its statehood in 1912 omitted it altogether, no small feat of obfuscation for a formation nearly one mile in diameter. Since the few impact studies being conducted at that time saw researchers dropping marbles into bowls of oatmeal and recording the ratio of displaced mush vis-à-vis the size of the offending cat's-eye, it is easy to understand why the crater's enormity and almost perfect roundness proved so baffling.

Since 1812, when a twelve-year-old girl named Mary Anning found a seventeen-foot-long *Ichthyosaurus* fossil along the cliffs of Dorset, humankind has been forced to come to terms with the upsetting evidence that many hundreds of thousands of startling and, sometimes, frightening genera came before it. The secondary recognition—that something caused these creatures to go extinct on a massive scale—soon followed. Mass extinctions are clearly evident in the fossil record and were duly noticed by nineteenth-century geologists (or "undergroundologists," as they were then known). They made these extinctions the basis of division (Cambrian, Devonian) within the geologic time scale. The question about mass extinctions was one of agency. Georges Cuvier, a brilliant French paleontologist and until Charles Darwin the most famous scientist of the nineteenth century, settled upon a "Doctrine of Catastrophes," which envisioned violent "revolutions" that all but swept clean ancient worlds of life. The events were "so stupendous," Cuvier wrote in 1821, that "the thread of Nature's operations was broken by them and her progress altered." Delighted Christians took this as proof of the Noachian deluge.

The opposing view of the world's prehistory was represented by the equally brilliant Scottish geologist Charles Lyell, who began as an admirer of Cuvier. Lyell believed that "we are not authorized in the infancy of our science, to recur to extraordinary agents." Lyell's view of the fate of species, most forcefully explained in his *Principles of Geology* (1830), was derived from a theory of slowly accumulating processes. Extinctions, then, were gradual affairs, and not subject to fantastical caprice. Lyell's view became known as uniformitarianism and proved so convincing that, within decades, to espouse any version of Cuvier's catastrophism became the scientific equivalent of wearing a tinfoil hat and claiming that streetlamps were issuing death rays. The gradualist interpretation of the world—which has, in most disciplines, served science well—came to dominate the study of mass extinction. In this century the demise of the dinosaurs has been associated with causes ranging from small ratlike mammals eating *Tyrannosaurus* eggs to gamma-ray bombardment from an exploding supernova to dinosaurs becoming too big to mount their partners to gradual climate change. (I remember vividly, as a young boy, weeping over an educational cartoon filmstrip that showed *Diplodocus* staggering through a desert and collapsing.)

But on June 30, 1908, something happened that would slowly begin to change the terms of this debate. Over the skies of a Siberian riverine area known as Tunguska (a word that has "9/11"-like resonance among astronomers), a small stony meteor no wider than sixty yards punched through the upper atmosphere and, due to aerodynamic pressure, detonated four and a half miles above the surface. The explosive force was that of 800 Hiroshimas and shook the Earth with the ferocity of a magnitude-eight earthquake. Seven hundred square miles of Russian woodlands were incinerated in seconds. That evening Europe's sky was so bright that there were reports of games of midnight cricket being played by uneasy Londoners. Amazingly, only two people were reported to have died in the Tunguska blast, as the area was mostly uninhabited. Had the asteroid been delayed by four hours, the explosion would have occurred over St. Petersburg (and, perhaps, prevented Soviet Communism). Russia was under considerable upheaval at the time—Czar Nicholas II had dissolved two successive parliaments in the preceding two years—and no one journeyed to Tunguska to investigate the event before 1927. Not until years after Hiroshima, Nagasaki, and the repeated irradiation of New Mexico and Kazakhstan were the physics of massive explosions properly understood, and the strangely craterless Tunguska site was subsequently adduced to have been caused by an asteroidal airburst, an event more common than one might suspect. The U.S. Defense Support Program's comprehensive satellite system, ostensibly used to provide for the global tracking of enemy bombers and missile launches, detects a dozen ten- to twenty-kiloton explosions in the upper atmosphere every year. A 1963 blast of 500 kilotons above Antarctica was initially mistaken for a nuclear test by South Africa. Until the popular emergence of UFO sightings, these explosions were commonly observed and reported by civilians.

Nonetheless, the danger posed by large-body impacts was discounted well into the 1950s, a time when a theory as elementary as continental drift was seeing its first acceptance. Nuclear weapons were still in their nativity, after all, and humankind had several thousand silos' worth of exterminating angels to worry over. By 1964 it was commonly believed that Earth possessed only six verified impact craters, and the impact-mass extinction link was ridiculed. In 1970 the Canadian paleontologist Digby McLaren went public with his belief that the possible culprit of the mass extinction at the end of the Devonian period, 365 million years ago, the fourth most intense mass extinction of all time, was impact-related. The Nobel laureate Harold Urey made another claim for impacts and mass extinctions in the journal *Nature*, in 1973, but since Urey was a chemist his research was thought suspect. Then, in 1980, Luis Alvarez, Walter Alvarez, Frank Asaro, and Helen Michel published in the journal *Science* an article entitled "Extraterrestrial Cause for the Cretaceous-Tertiary Extinction."

The Cretaceous-Tertiary extinction, which occurs at what is known as the K-T boundary (C having already been secured by the Carboniferous period), marks the sixty-five-million-year-old point at which the dinosaurs go AWOL from the fossil record. Even at their acme of diversity (small children should probably stop reading now) no more than fifty species of dinosaurs, a decidedly trivial portion of Mesozoic Era life, were alive at one time. At the K-T boundary, as few as twenty-five saurian species were left. None survived past it. Of the mammalian species, perhaps ten or fifteen survived the K-T mass extinction, commonly thought to be the second most profound of its kind. Similar losses are mirrored in the fossil record of every species alive at the time.

The Alvarez group determined that the K-T boundary clay had an anomalously high incidence of the element iridium. Since iridium is roughly 5,000 times more abundant in extraterrestrial objects than in Earth's accessible crust, it seemed clear that something cataclysmically extraterrestrial in origin occurred at the K-T boundary, an event that showered Earth with the element. These findings were attacked for several years, leading the *New York Times*, in 1985, to issue a now famously dyspeptic editorial. "Astronomers," the *Times* scolded, "should leave to astrologers the task of seeking the cause of earthly events in the stars." (Stephen Jay Gould brilliantly ridiculed the *Times* by writing a fictitious editorial dated 1633: "Now that Signor Galileo . . . has renounced his heretical belief in the earth's motion, perhaps students of physics will . . . leave the solution of cosmological problems to those learned in the infallible sacred texts.") Not until the discovery, several years later, that the "shock metamorphism" of large-body impacts (and nuclear explosions) can form two separate minerals, stishovite and coesite, and that many suspected impact sites had high concentrations of both, did the hypothesis begin to win converts. A mid-1980s poll revealed that 90 percent of American scientists quizzed accepted that impacts occurred, though only 4 percent accepted an impact as the explanation for the K-T mass extinction. Previous beliefs that large-body impacts affected only the "lethal area" in the direct vicinity of the strike were, however gradually, abandoned. Impacts may happen, it was at last conceded, but they were freakish, remote events that did not script the fate of biology.

In 1990 the eminent late astronomer Eugene Shoemaker presented a paper to the Geological Society of America that demonstrated, as never before, the sheer number of Earth-crossing asteroids in our solar system, concluding that "Earth resides in an asteroid swarm." With these six words, Shoemaker finally, viscerally stated what scientists had been willing to accept only with decorous academic distance: millions of pieces of rock, some of them massive, were flying past Earth at staggering speeds, and sometimes these rocks hit us. All we had to do was have a look around at our beat-up planet for the well-documented proof. Opponents of the impact hypothesis suffered further attrition when, under the tip of the Yucatán Peninsula, near the Mexican port city of Progreso, a sixty-five-million-year-old crater 120 miles in diameter was linked decisively to the K-T mass extinction. The crater, called Chicxulub, had been found as long ago as 1978, but because its discoverer, an oil-industry geologist named Glen Penfield, was not an academic, converts were hard-won. By the early 1990s Chicxulub had become widely accepted as the scar of a large-body impact that finished off the dinosaurs and 75 percent of all other life on Earth.

Currently almost 1,000 potentially civilization-ending NEAs are known to have orbits passing within five million miles of Earth. It is statistically likely that most of us will live to see another impact, however small or large. In 1996, asteroid 1996 JA1 came within 200,000 miles of hitting Earth. We were provided with three days' notice. A week later, asteroid 1996 JG came within 2,000,000 miles of hitting Earth. We did not even see it until it had already passed. Both were twice as large as the Tunguska asteroid, and either could have killed one percent of Earth's human population; that is, 60 million people.

With the acceptance of the impact hypothesis has come another, related area of study: the frighteningly prominent role smaller impacts are now thought by some to have played in known human history. Asteroids themselves have traditionally been subject to great, if puzzled, veneration. In ancient China, for instance, people used to grind up and eat meteorites, and other cultures took the metal from M-type meteorites and smithed them into weapons. Simply put, it is very likely that smaller catastrophic impacts have occurred far more often than most people realize. A 1,000-megaton event in Argentina resulted in mile-wide craters not thought to be more than about 10,000 years old. A small (not quite mass) extinction occurred 10,000 years ago, which finished off, primarily in the Americas, larger terrestrial mammals such as the woolly mammoth and the saber-toothed tiger. If we project forward another 6,000 years, we watch several fairly advanced civilizations, such as that of Ur, fall and vanish. We find that bookkeeping of the skies becomes, quite suddenly, a matter of some cultural importance. The literature of the ancient cultures that proceeded from Ur, including that which became the Judeo-Christian, contains much lore of a vast flood, lost continents, vanished oceans, a world gone meteorologically mad. We find Egypt, which traced its origins to Pythom, apparently a falling object of some kind, and we find a strange and seemingly offhanded Egyptian memory, now forever enshrined in Exodus, of a darkness lasting three days. We find an ancient city in Arabia, Wabar, supposedly destroyed by another falling object, an event that happens to coincide with the rise of Babylonian astrology. There are counterexplanations to all of these developments, of course, just as heaven is a counterargument to death—it exists primarily because we need it to. If it is true that small though still catastrophic impacts—caused by asteroids not big enough to find with existing technology—are statistically timed to occur roughly every few thousand years, this means that the woolly mammoth and Ur fell victim to separate impacts, and it means that another is due, well, any day now.

Unlike ancient human beings, however, we actually can do something about this other than devise cosmologies of desperate emotional necessity. There are several theoretical defenses to asteroid impacts. One method involves landing a rocket-booster device on the collision-course asteroid and then attempting to steer it from Earth's path. Another proposes placing solar panels on an incoming asteroid to create what is called a "solar-powered mass driver," thereby altering its trajectory in a fuelless, less-expensive way. Yet another envisions painting asteroids black in the hopes that the change in absorbed solar radiation would gradually shift the object's orbit. But the most imperative theories involving asteroid defense are nuclear. It seems beautifully, if demonically, apt that the weapons that have terrified us all for half a century might turn out to possess the cleansing holy fire we were all promised they did. How effective they would be against the unideological phenomenon of asteroids is another matter. Simply hitting the rocks with missiles would accomplish nothing; they are traveling too fast, and the warheads' tonnage would only be absorbed. The most useful deployment of nuclear force against an unopposable asteroidal object would be to explode a powerful nuke at the asteroid's surface near the closest point its orbit takes to the sun, where the leverage of deflection is greatest. Then our planet would wait. If this first explosion did not alter the asteroid's course enough (chances are it would not), one would try another explosion, and perhaps another, thus "herding" the asteroid into a different orbit. Problems with this method, and there are many, include the possibility that the asteroid in question has been weakened by an older collision, or is loosely bound together (a rubble pile, this is called), or has an insecure interior riddled with open space. Nuclear weapons, used against such unstable bodies, might only transform one asteroid coming our way into several asteroids, perhaps hundreds—pieces of which would be large enough to breach Earth's atmosphere. The other problem would be an enlivened nuclear-weapons industry, one busily developing warheads not as a deterrent but as devices *intended* to be used. No one would object, of course, if we were to keep a few of these newer-line missiles—which would, no doubt, be better than ever—on high alert, or deployed a few more here or there. There are maniacs out there. Maniacs with nuclear weapons! China, for its part, has explained its resistance to the anti-nuclear-proliferation Comprehensive Test Ban Treaty on the grounds that it would like to keep its missiles in the event of an asteroid-impact threat. No doubt Iran and Iraq and Libya and Algeria would like to step up their own programs, too. Just in case.

Helpless

READING ABOUT ASTEROID IMPACTS will undoubtedly cause many people distress. I feel bad about that, and I would like to say that although these threats are terrifying, all is not lost. Concerned, dedicated people are working on the asteroid-impact threat, and one need not be a deluded idealist to believe that they may succeed. Hope, after all, takes as its foundation not likelihood but possibility. There is, however, another threat to ourselves and our civilization, one that cannot be stopped or avoided. You readers who find yourselves already traumatized, let me entreat you here, please, to stop reading.

Comets differ from asteroids in several ways. Consensus holds that they are "dirty snowballs" made up of ice and carbon-bearing rock. A 1986 "flyby" mission to Halley's comet beamed back data suggesting that its nucleus is less dense than water, 50 percent of it a warren of cracked and empty networks. How this pertains to other comets is not known. Before the comet Hyakutake was found in 1996, only five previous comets had been detected, by radar. Traveling like frozen freight trains along the loneliest edges of the solar system, comets occasionally enter the inner solar system, the neighborhood of Earth, at twenty-six miles a second, leaving behind them a long tail of dust and gas crystals that can stretch back as far as sixty million miles. Replacement dust accumulates on cometary surfaces; when the dust layer becomes thick enough, comets gain an excellent shield against solar heating. An icy skein builds, and they get bigger. One cometlike object, Quaoar, recently found floating out around Pluto, is 800 miles across. As comets approach the sun, however, their frozen gases expand and form makeshift jets that can alter their course. This makes predicting accurate orbits of newly discovered comets nearly impossible. But discovering them is also challenging. They are too fast, not typically seen until they pass near the sun, and in any event their gas- and dust-obscured passage across the universe's dark starry backfield is often difficult to discern.

We know of two types of comets. The closest to Earth, called short-period comets, are found just beyond Neptune in the Kuiper Belt, named after the astronomer Gerard Kuiper. Long-period comets make up what is known as the Oort cloud—named in honor of the astronomer Jan Oort, who first hypothesized its existence—an envelope of as many as a trillion comets that travel around the sun far beyond Pluto. The Kuiper Belt and Oort cloud are both remnants of the early physical conditions of the primitive solar nebula. The Oort cloud was most likely formed by gravitational ejecta from the Uranus and Neptune regions, and some astronomers believe that the Kuiper Belt is merely the innermost edge of the Oort cloud. Jupiter's massive gravitational force shields Earth from many long-period comets, but its movement is also thought to be the key mechanism for injecting the few comets that manage to get past it, often in shattered form, into short-period Earth-crossing orbits. Almost all long-period comets have orbital periods of 100,000 years or more, making it all but certain that there are literally millions of comets with periodic near-Earth passes we know nothing about. We could have as little as three months' warning when a comet on a collision course with Earth appears in the sky. Most are too big to stop with nuclear weapons, which does not much matter, as their meddled-with, chaotic trajectories make intercepting them fantasy.

Edmund Halley, in his *A Synopsis of the Astronomy of Comets* (1705), was the first scientist to wonder if comets might impact Earth. Halley's astonishing foresight was only negligibly explored until the 1970s. In the 1990s, however, comets stepped to the forefront of scientific thinking about apocalyptic mass extinction. On March 24, 1993, three years after concluding that the "Earth resides in an asteroid swarm," Eugene Shoemaker, along with his wife, Carolyn, and the amateur (though highly respected) astronomer David Levy, detected upon a photographic plate a strange smear of pearly light. What they had found was twenty-one pieces of a comet that had been torn apart by Jupiter's orbit the previous July. The pieces of this comet, now called Shoemaker-Levy 9, were predicted to impact the planet the following year. Sadly, the impact would occur on Jupiter's dark side, viewable only at a great distance by the Voyager and Galileo probes. Many worried that humankind's first opportunity to observe a large-body impact in at least 800 years (or since what may have been the formation of the moon's Giordano Bruno crater, in 1178) would be spoiled.

They needn't have worried. Upon impact, the larger pieces of Shoemaker-

Levy blew fireballs thousands of kilometers high into Jupiter's atmosphere. When Fragment G collided with the King of Planets two days after the first impact, the flash was so bright that sensitive instruments trying to measure it were momentarily fried. The temporary scar left by Fragment G was larger than Earth itself, and the explosive energy released was the equivalent of a Hiroshima-sized nuclear bomb exploding every second for thirteen years. Shoemaker-Levy's volatile swan song was quickly and accurately called the astronomical event of the century, and has proved the starkest challenge so far to humanity's sense of its own inviolability.

Scientists are divided on whether the K-T mass extinction sixty-five million years ago was caused by a comet or an asteroid. The severity of the event—miles of evaporated ocean; the very high chance that a hundred trillion tons of molten rock were thrown into space, frozen, and then pulled back down to the surface of Earth in the form of more impacting meteorites; an ozone layer so shredded that any creature peeking out of its cave even a year after the impact would have found its skin on fire in the ultraviolet spring; the sheer number of extinctions—points to a comet. It is empirically inarguable that every few dozen million years a mass extinction is visited upon Earth. Various arguments place these mass extinctions, the extent and agencies of which are still debated, at intervals ranging from twenty-six to thirty million years. A theory called the "Shiva Hypothesis," named after the Hindu god of destruction, holds that mass extinctions occur in startling simultaneity with the movement of our solar system through the galactic plane, a passage that is thought to perturb millions of Oort cloud comets into our path. Comets, and the mass extinctions they cause, might very well be the piston that drives Earth's biological processes. If the Shiva Hypothesis is correct, we are all just marking time until the next comet arrives.

BEFORE LEAVING THE JET PROPULSION LABORATORY, I stopped to visit Don Yeomans, a man commonly regarded as one of the world's key figures in near-Earth object studies, to talk about comets. Yeomans struck me as a former nerd who has aged extremely well. He radiated a thoughtful, deceptively low-key intelligence. That he is well liked was clear from the boxed Yeomans action figure someone had given him, and from the fact that the writers of a doomsday television movie named their impacting comet after him. When he went through his files to find the film's title, I noticed that one of his drawers was labeled "NEOs and Things That Go Bump in the Night . . ."

"We're hit by tons of material every day," Yeomans told me after his futile file search, "but it's all dust. We're all walking around with comet dust in our hair. What really interests me is that comets and their impacts with Earth are—apart from the sun—the only bodies that have a direct effect on evolution, on life, on the origin of life. They probably brought the materials of life to Earth—carbon-based molecules and water. They're not just something in the sky that looks nice, like the outer planets or stars or clusters. That's all very interesting but has no real effect on us. When you come right down to it, the guy in the street wants to know, 'What's in it for me? Why should we be studying these things?' And comets, I always claim, pass the brother-in-law test."

"The brother-in-law test?"

Yeomans smiled. "My brother-in-law doesn't believe in the space program that much. And he says, 'Hey, why are you guys spending millions of dollars to do this and that?' But after I explained comets and the origin of life and evolution, he said, 'Maybe that makes sense.' "

Yeomans was addressing the notion that impacts upon the ancient atmosphere, before the origin of life on Earth, generated a stew of simple organic molecules that eventually resulted in amino acids and nitrogen bases able to serve as the keystones of life. Impacts are high-energy processes not unlike ultraviolet light, cosmic-ray irradiation, and lightning discharges. The atmospheric violence caused by impacts can also create a high incidence of these potentially life-creating processes—a case of God appearing in the whirlwind, hurricane, earthquake, and tsunami.

Human beings are not a goal that this stew evolutionarily pushed toward; they are only one entity within evolution. Contrary to the worries of certain Kansas school boards, evolution does not, in fact, push toward *anything*. The Darwinian evolution attacked by creationists is not even Darwinian but an inflexible version of natural selection promoted by Darwin's followers that Darwin himself would not have recognized. Today, Darwinian natural selection is an acknowledged fact of life's micro development, but on a macro scale natural selection is moot. How does an organism adapt when its world is hit without warning by a ten-mile-wide ball of ice traveling 70,000 miles per hour? Evolution is not even good for the planet, based as it is on speciation and phyletic transformation. If these processes were allowed to go on indefinitely the planet would be overrun and all life would go extinct. Thus mass extinction serves as speciation's necessary foil, and the victims run well into the millions: today's extant species account for less than one percent of the total number to have ever lived. "Mass extinctions," wrote Stephen Jay Gould, "can derail, undo, and reorient whatever might be accumulating during the 'normal' times" between "imparting a distinctive, and perhaps controlling, signature to diversity and disparity in the history of life."

"These objects are weird," Yeomans concluded. "And the history of these objects is weirder still. Comets are unlike anything else. They show up unexpectedly, they disappear unexpectedly, they have different shapes and sizes and characteristics. There's nothing celestial-looking about them. They are the wild cards."

By the time I left Yeoman's office, The End of the World and its wild-card causes had begun to seem less like a force of impartial terror than something far more complicated—something *necessary*. Life is a huge blackboard filled with a million marks of chalk. Every thirty million years that chalkboard is forcefully wiped clean, leaving only a few small smudges in the corners, whereupon life begins again without regard to perfection of adaptation, what has come before it, or the miserable consciousnesses of those few creatures able to wonder why they are here. For reasons no one yet understands, smaller animals seem to have a survivalist edge in the aftermath of most mass extinctions. One such (most likely shitlessly frightened) animal, a primate, survived the K-T mass extinction. Possibly it lived in a wet, boggy area, the only part of the planet that would not have burned up in the global firestorm. Say its name: *Purgatorius.* We may have this tiny creature to thank for our current civilization. So impacting comets giveth, clearing the way for new species, and they taketh away. This most unstoppably terrifying vision of the world's end is also the most comforting, for it forms an iron law no organism, including ours, can hope to step around. Life cannot win. "Behold," the Lord says in Isaiah 65:17, "I create new heavens and a new earth: and the former shall not be remembered, nor come into mind." There is freedom in that recognition, a sense of knowing where we stand in fighting to stave off oblivion, and of knowing where we surrender. As everything must.

ELIZABETH ROYTE

Transsexual Frogs

FROM *DISCOVER*

Tyrone Hayes stands out in the overwhelmingly white field of biology, and his skin color isn't the half of it. To use his own idiom, Hayes is several standard deviations from the norm. At the University of California at Berkeley, he glides around his lab wearing nylon shorts and rubber flip-flops, with a gold hoop in one ear and his beard braided into two impish points. Not counting his four inches of thick, upstanding hair, Hayes is just over five feet tall, with smooth features and warm eyes. He drives a truck littered with detritus human, amphibian, and reptilian. He keeps his pocket money in a baby's sock. "Hey, wassup?" he'll say to anyone, from the president of the United States on down. He can't help the informality, he says. "Tyrone can only be Tyrone."

Hayes, 35, is a professor at Berkeley, where he has taught human endocrinology since 1994. His research centers on frogs, of which he keeps enormous colonies. Frogs make convenient study subjects for anyone interested in how hormones affect physical development. Their transformation from egg to tadpole to adult is rapid, and it's visible to the naked eye. With their permeable skin, frogs are especially vulnerable to environmental factors such as solar radiation or herbicides. That vulnerability has lately garnered Hayes more attention than his appearance ever has.

The controversy began in 1998, when a company called Syngenta asked

Hayes to run safety tests on its product atrazine. Syngenta is the world's largest agribusiness company, with $6.3 billion in sales of crop-related chemicals and other products in 2001 alone. Atrazine is the most widely used weed killer in the United States. To test its safety, Hayes put trace amounts of the compound in the water tanks in which he raised African clawed frogs. When the frogs were fully grown, they appeared normal. But when Hayes looked closer, he found problems. Some male frogs had developed multiple sex organs, and some had both ovaries and testes. There were also males with shrunken larynxes, a crippling handicap for a frog intent on mating. The atrazine apparently created hermaphrodites at a concentration one-thirtieth the safe level set by the Environmental Protection Agency for drinking water.

The next summer Hayes headed into the field. He loaded a refrigerated 18-wheel truck with 500 half-gallon buckets and drove east, followed by his students. He parked near an Indiana farm, a Wyoming river, and a Utah pond, filled his buckets with 18,000 pounds of water, and then turned his rig back toward Berkeley. He thawed the frozen water, poured it into hundreds of individual tanks, and dropped in thousands of leopard-frog eggs collected en route. To find out if frogs in the wild showed hermaphroditism, Hayes dissected juveniles from numerous sites. To see if frogs were vulnerable as adults, and if the effects were reversible, he exposed them to atrazine at different stages of their development.

Hayes published his first set of findings in April, 2002, in the *Proceedings of the National Academy of Sciences.* He published the second set in October, in *Nature.* Both times the media went a little crazy. The two studies showed equally dramatic results: 40 percent of male frogs were feminized; 80 percent had diminished larynxes. Wild frogs collected from areas with atrazine showed the same number of abnormalities. Could the chemical also affect humans? The beginning of an answer may be emerging. Workers at a Louisiana plant where atrazine is manufactured are now suing their employer, saying they were nine times as likely to get prostate cancer as the average Louisianan.

INSIDE BERKELEY'S VALLEY LIFE SCIENCES BUILDING, Hayes approaches a set of double doors and lifts his thigh, doggy style, toward the wall. The doors respond to a security card in his pocket and swing wide onto an empty corridor. It's 7 AM, but Hayes has been here since 4:30 this morning, when he came to "make water"—mix the chemical cocktails in which he's raising 3,000 leopard frogs in a crowded basement lab. He deftly shakes crickets— frog breakfast—from a plastic bag into dozens of tanks. On another shelf, tadpoles swim in one set of deli cups while metamorphs, which have both tails and legs, swim in another. Escaped crickets dart around the room. Strips of colored tape adorn each tank, each color denoting a particular mix of compounds. In this quadruple blind experiment, neither Hayes nor his assistants know exactly what they're testing. Except for the notorious Red Yellow Red.

We peek into the suspect tank. "They're not doing too well, are they?" Hayes says, brushing a cricket off his neck with a practiced flick. The frogs are listless. Their heads tilt at a creepy angle. "Everything we put in this mixture died within a week, except for frogs that have adapted to that environment. So I had to look it up." Red Yellow Red, the codebook said, is the brew that runs off a Nebraska cornfield in springtime. "These frogs took a month longer than average to metamorphose, and then they were smaller than average," Hayes says. "That's wrong: Usually a longer metamorphosis means a bigger frog." He dumps in another meal of crickets and delivers the kicker: "This mixture from the cornfield has a lower dose of chemicals than what's in the drinking water there."

The problem, Hayes knows, goes well beyond frogs that loiter near cornfields. According to James Hanken, a biologist at Harvard University who heads a task force on declining amphibian populations, "at least one-third to one-half of all living species of amphibian that have been examined in this regard are on their way down, and out." Researchers have offered a number of explanations for the die-off: attacks of parasites, exposure to radiation or ultraviolet light, fungal infections, climate change, habitat loss, competition with exotic species, and pesticides. Atrazine is used in more than 80 countries, primarily on corn and sorghum fields. By interfering with frog reproduction, Hayes wonders, could it be part of the problem?

Atrazine is a synthetic chemical that belongs to the triazine class of herbicides. Its technical name is 2-chloro-4-ethylamino-6-isopropylamine-1,3,5-triazine. In the United States, farmers apply around 60 million pounds of atrazine a year. Nearly all of it eventually degrades in the environment, but usually not before it's reapplied. The EPA permits up to three parts per billion of atrazine in drinking water. Every year, as waters drain down the Mississippi River basin, they accumulate 1.2 million pounds of atrazine before reaching the Gulf of Mexico.

Like the smoke from factory chimneys, pesticides cross borders. Atrazine molecules easily attach to dust particles: Researchers have found it in clouds, fog, and snow. In Iowa the herbicide has been documented at 40 parts per billion in rainwater. According to the U.S. Geological Survey, atrazine contaminates well water and groundwater in states where the compound isn't even used. "It's hard to find an atrazine-free environment," Hayes says. In Switzerland, where it is banned, atrazine occurs at one part per billion, even in the Alps. Hayes says that's still enough to turn some male frogs into females.

Hayes talks rapidly as he walks from the basement lab. He'll also talk rapidly as he drives to his children's school in an hour, as he eats at a nearby restaurant, and as he types e-mail. "I'll calm down after lunch," he promises. "Here's how we think it works. Testosterone is a precursor to estrogen. In male frogs, it makes their voice boxes grow and their vocal sacs develop. But atrazine, in frogs, switches on a gene that makes the enzyme aromatase, which turns testosterone to estrogen. Normally, males don't make aromatase; it's silent. In these males, the estrogen induces the growth of ovaries, eggs, and yolk." We're at the double doors, and Hayes lifts his thigh again. "So you've got two things happening: The frog is demasculinized, and it's also feminized."

And the females that get extra estrogen? "It wouldn't happen," Hayes says. "There's a feedback mechanism. The excess hormone would decrease stimulation of the ovary, which would then cut off its production of estrogen."

Because hormones, not genes, regulate the structure of reproductive organs, vertebrates are particularly vulnerable to their environment during early development. Frogs are most susceptible just before they metamorphose. Unfortunately, that change occurs in the spring, when atrazine levels peak in waterways. "All it takes is a single application to affect the frog's development," Hayes says.

Theo Colborn, a senior scientist with the World Wildlife Fund who has spent nearly 15 years studying endocrine-disrupting chemicals in the environment, calls Hayes's work a breakthrough. "At a time when other developmental biologists were taking a broad, traditional approach, he was taking long-term effects into consideration," she says. "No one had looked at the histology the way he has. Everyone was so hung up on limb deformities in frogs that they forgot about other effects. His work may explain why frogs are disappearing."

HAYES HAS ALWAYS BEEN fond of frogs. He grew up in a modest neighborhood of brick houses outside Columbia, South Carolina. The development had been drained of its marsh, but snakes, turtles, and amphibians abounded. Hayes followed them and learned their ways. As a teenager, he dug a pond in his backyard, hoping to breed turtles. He kept lizards. His father brought him boxes of *National Geographic* from houses in which he had installed carpet. The boy read them all. "Those magazines were the beginning of it," Romeo

Hayes says. "Even then he knew he wanted to be a scientist." The television was always on in the Hayes household, even during meals, and Tyrone paid particular attention to the nature specials. When he began dating, he took girlfriends to the Congaree Swamp, nine miles away. The young women assumed he had other things in mind, but his motives were always the same: He wanted help catching frogs.

The summer after sixth grade, Hayes taught himself to play basketball. "That was the only way I knew for blacks to get into college," he says. Through high school he wrestled, struggling with a hypothyroid condition to make weight. Entranced by the pop star Prince, he wore frilly shirts and velvet jackets, winning Best-Dressed Student five years in a row. "I wanted a hoop earring, but my mother forbade it," he says. Within days of arriving at college, he pierced his ear himself. (These days, Hayes wears a coat and tie to meetings. "But it's a real *Men in Black* kind of suit," one former student says. "And he wears a skullcap.")

Geography and family circumstances narrowed Hayes's expectations. His father had been the first on his side of the family to attend high school. Tyrone had never heard of an academic scholarship; he had never known anyone who left South Carolina to go to school. But his high PSAT scores brought a sheaf of recruitment letters to his house. He wrote a personal statement about his interest in armadillo biology and mailed it to Harvard. It was the only school to which he applied. "I'd heard of it on *Green Acres* and figured it must be good," he says, without a trace of irony.

Once on scholarship in Cambridge, Hayes thought he'd become a doctor. Then he began working with the biologist Bruce Waldman on kin recognition in toads. Waldman recognized Hayes's talent for asking challenging research questions and his skill in the field and the lab. He treated the freshman like a grad student. Soon Hayes was studying environmental effects on tadpole metamorphosis. "I realized what a person who enjoyed what I did might do for a living," Hayes says. "I saw the whole picture coming together."

Still, nothing in his background had prepared him for Harvard's social and academic pressures. "Most blacks at Harvard were from private schools," Hayes says. "They knew what was going on. Their parents had gone to school there. They flew to Bermuda at spring break." Hayes felt out of place. He didn't join any campus groups and spent all his time in the laboratory. "It was the only place I felt at home," he says. "I had four finals to study for and didn't know how to organize my time. I couldn't get advice from my dad." His grades fell, and he was placed on academic probation. Hayes nearly dropped out at that point, but Waldman and Kathy Kim, the girlfriend he later married, persuaded him to stick it out. In 1989 he graduated with departmental honors and moved to Berkeley, where he earned his PhD at the age of 24.

"You think Tyrone is manic now, you should have seen him in those years," says Nigel Noriega, a research scientist in reproductive toxicology at the EPA. At Berkeley, Hayes's weight ballooned from 135 pounds to 260 pounds in six months. To get back into fighting shape, he ran 18 miles a day, often with an infant in a stroller. He went for days without sleep, then set the alarm to ring after just a few minutes. He was running a shape-shifting experiment on himself.

He drove his students to the edge as well. Lab assistants, drawn in by his dynamism, became exhausted and depressed. "It was hard; we barely saw the light of day," says Roger Liu, who spent the better part of 10 years in Hayes's lab. The results of their experiments would be so far in the future that they lost sight of their goals. Still, they loved Hayes. "Tyrone treated undergrads like grad students and grad students like postdocs," Noriega says, echoing Hayes's assessment of Waldman. "You could ask him for anything." When Hayes found attendance flagging at his 6:30 AM lab meetings, he started baking, at 2 AM, to lure students in. When he worried about his charges walking to the lab in the dark, he picked them up at 4 in the morning, shining a spotlight into their windows to wake them.

From the outset, Hayes's lab attracted minority students. It soon became far and away the most diverse in the department of integrative biology, which is only 3 percent black and has produced just four black PhD's in its history. (Noriega is one.) Now nearly 20 percent of his lecture class is black. Hayes says he concentrates on selecting talented students who need nurturing. This semester's crop of researchers comes from Vietnam, India, Pakistan, Thailand, Tunisia, Mexico, Guatemala, Canada, and the United States.

"Maybe minority students think they'll make some kind of connection with me," Hayes says, shrugging. Or maybe they appreciate his holistic approach to science. He often brings his two children—Tyler, 10, and Kassina, 7—into the lab with him, and he watches over his students with the same paternal eye. "The lab was like a family," Liu says. "Dad got pissed, the siblings fought, but we were happy."

Last year, at the departmental graduation, the students gave Hayes a standing ovation. This past spring he won the College of Letters and Science's award for Distinguished Research Mentoring; a week later he won its Distinguished Teaching Award. "Tyrone reveals that science is inbred and flawed and political, just like art and music," Noriega says. "But he's still striving for its bright and shining truth. He lays all this out, and you see it's still worth it."

Even after all the weirdness with Syngenta.

LIKE ALL CHEMICAL COMPANIES, Syngenta has to have its products tested for safety before the EPA will approve them. The company came to Hayes in 1997 because he had experience with hormones and amphibians: He had developed an assay in which frogs exposed to estrogen mimics turned from green to red. "This was a chance to use my research," Hayes says. "Also, not that many labs are set up to travel and collect eggs, establish a colony, and breed. I had a big lab, with lots of people willing to move 3,000 frogs from tanks to deli cups."

Hayes says that when he informed Syngenta about atrazine's negative low-dose effects in August 2001, the company treated his data like a hot potato. "They told me, 'That's not what you were contracted to do. We don't acknowledge your work,'" Hayes says. "I sent them all my raw data, and they FedExed it back to me." Ronald Kendall, an environmental toxicologist at Texas Tech University and a leader of Syngenta's atrazine-testing panel, insists that Hayes told the team only about the frogs' shrunken larynxes, not their hermaphroditism: "We didn't learn about gonadal effects until a hormone meeting late in November." Rather than keep quiet about his findings, Hayes quit his contract and repeated his experiments. The week before he was scheduled to share his data with the EPA, he received 500 computer viruses.

With Hayes no longer on board, Syngenta funded some of Kendall's colleagues at Texas Tech to replicate the work. They produced almost no hermaphrodites at the atrazine levels Hayes had tested. The lab conditions in Texas differed from conditions in the Berkeley basement. For example, the Texas experimenters raised their frogs in glass instead of plastic tanks, at higher population densities, and at cooler temperatures, and they fed them differently. "But if the effect is robust, as Hayes claims it is, you should still be able to see it under slightly different conditions," says James Carr, a comparative endocrinologist on the Texas Tech team.

Hayes accused the Texas team of raising unhealthy frogs in tanks with uncontrolled atrazine levels. "Their animals were underfed and overcrowded," he says. "How can you tell if their gonads are deformed if the animals don't develop properly?" In response, the Texas team crafted an 18-page defense, to which Hayes responded with 22 pages of his own. The Texas team says it was difficult to compare the health of their animals with the health of Hayes's because he didn't report hatching success, mortality, survivorship, and other data. Hayes responds: "They've had all my information on protocols and SOPs since 1999. They signed off on this work. They even visited my lab."

While the scientists sparred, workers at Syngenta's atrazine plant in St. Gabriel, Louisiana, stole the spotlight when their cancer rates became public. At least 14 of 600 employees who'd been at the plant for more than 10 years had developed prostate cancer—a rate nine times as high as that of the general statewide population.

Had Syngenta inadvertently tested atrazine on humans? Studies of farm laborers who worked with the compound showed rates of certain cancers double to eight times the national average, but those exposures were intermittent and not exclusive, because workers handle many types of chemicals. In St. Gabriel, atrazine represented 80 percent of the plant's production, and it was made year-round. Atrazine dust covered the walls and floors, countertops and lunch tables.

Hayes's frog data were alarming, but they probably wouldn't have persuaded the EPA to ban atrazine. The cancer findings may. This past summer, the Natural Resources Defense Council persuaded the agency to launch a criminal investigation of Syngenta for suppressing data on the herbicide's potential risks to the environment and to human health. The EPA has since extended the deadline for its atrazine review. In addition to the cases in Louisiana, laboratory studies have linked atrazine to hormonally responsive cancers in humans and lab animals. Studies have also suggested that it disrupts the production of hormones such as testosterone, prolactin (which stimulates the production of breast milk), progesterone, estrogen, and the thyroid hormones that regulate metabolism.

Nonetheless, Hayes doesn't jump to condemn atrazine. He says he hasn't studied humans, but it is unlikely they'd be affected because atrazine doesn't accumulate in tissues the way DDT does. Others aren't so sure. "Why would anyone think these pesticides *wouldn't* affect us?" the World Wildlife Fund's Theo Colborn says. "No matter the species, we all have similar signaling systems in our bodies, similar chemical reactions. That's why we've always tested drugs on animals." Human kidneys filter atrazine, and humans don't spend a lot of time swimming in pesticide-laced water, the way frogs do. But human fetuses do live in water.

"Our big concern is pregnant females," Colborn says. "There have been enough studies on farm families to show that babies conceived in the spring, when runoff is highest, have far higher rates of birth defects than babies conceived at other times." But what component of the runoff is toxic and at what levels? That may be impossible to say, because scientists don't run lethal-dose experiments on humans. Faced with this uncertainty, how cautious should we be? When pressed, Hayes says that if his wife was pregnant, he'd advise her against drinking water from much of the Midwest—his children too. "If there's a .01 percent freak chance that something could happen, why take that chance?"

THE MYSTERY OF AMPHIBIAN DECLINE continues to intrigue Hayes. He believes a combination of many different effects may stress frogs' immune systems and that atrazine may be a part of it.

He dreams of testing his ideas with the perfect field experiment, one without unquantifiable variables, and he knows just where he'd enact it. "We'd go to Biosphere 2, in Arizona," he says. "We'd bring in all our own air, our own water. We'd set up farm plots with corn. We'd bring in our own frogs, study every compound and its impact on the corn, the corn pests, the nontarget organisms." There's a gleam in Hayes's eyes. The thought of all those animals, the long hours, the phalanx of tired graduate students—it all makes his blood rise. "Nobody knows what these compounds actually do," he says. "I want to figure it out from beginning to end."

SUSAN MILIUS

Leashing the Rattlesnake

FROM *SCIENCE NEWS*

Depending on how you look at them, snakes have no neck or nothing but neck, and either way, Ron Swaisgood had a problem. To finish his PhD at the University of California, Davis, he had to figure out how to put a rattlesnake on a leash. Obviously, dropping a slipknot around the snake's neck wouldn't do. Swaisgood's research project required that the snake comfortably slither, coil, and strike but still be tethered tightly enough that there was no chance it could escape.

Swaisgood eventually developed a great snake leash, finished his degree, and proceeded to his current job at the San Diego Zoo. The scientific paper based on the research just says he tethered the snake and then goes on to describe the ways that ground squirrels assess snakes as threats. The leash joined thousands of other little unsung triumphs of creativity in experimental design that don't usually show up in scientific articles—or reports in *Science News*. Ask for the details, though, and another side of research appears. Here, many biologists say, is a lot of the fun.

Experimental science demands creativity in solving problems great and small, and the study of animal behavior makes a dramatic showcase for the process. Animals clearly have minds of their own, and to elucidate such foreign perspectives, researchers have to design cleverly, from the basic strategy of the experiment to dozens of lesser details. The traditions of animal behavior celebrate do-it-yourself flair, and experiments mix the sublimely high tech with the ridiculously simple. Consider the snake leash.

Swaisgood reminisces that the first Northern Pacific rattlesnake he'd caught for his experiment "was kind of a phlegmatic snake." The university veterinary surgeons who frequently prepare animals for experiments gave the rattler an implant of a little plastic loop. Swaisgood then just clipped a line to the loop and staked the snake in place.

The next snake, however, had a completely different personality and moved vigorously. A loop implant didn't look like a good idea. "We were concerned he might hurt himself," Swaisgood says. "We started going through our tackle box of research gear looking for something else." He finally leashed his rattlesnake by attaching fishing line to it with—bonus points to readers who saw this coming—that icon of invention: duct tape.

THE CORE OF AN EXPERIMENT in behavioral science depends on changing one thing about an animal's world but leaving everything else the same. This challenge sometimes demands a mix of scientific sophistication and parlor tricks.

Ken Kardong couldn't use a silk handkerchief as a blindfold since the eyes he needed to cover belonged to a rattlesnake. After Kardong's lab at Washington State University in Pullman had documented the biomechanical marvels of the snake's strike, he began thinking that the snake's sensory systems needed to be equally marvelous to take advantage of that natural engineering. He decided to study how the snake's senses contributed to strike targeting, so he wanted to block, and then restore, each of its senses.

Snakes' eyes carry a clear, protective layer that lacks the nerve endings that make human eyeballs so tender. "It's like they're wearing glasses," says Kardong. So, with a conventional snake-handling stick, he held a rattler down on a counter, got a firm grip at the back of its head, and placed a patch of black electrical tape over each eye.

Kardong next had to invent another kind of blindfold, this one for the infrared-sensing pits on the snake's face. Depending on the size of the snake, the pair of pockets varies from depressions the size of a pinhead to pits that could hold a BB.

Again, Kardong thought of electrical tape, but he wanted extra insulation to block heat. At first, he considered rolling up little balls of paper. Then, a brainstorm came from a Styrofoam coffee cup. He sliced it into strips and rolled bits between his thumb and forefinger until he had the right size for each snake.

Kardong says, "My definition of cleverness includes 'simple.' "

Blindfolding either the eyes or the pits, by the way, made no difference in the accuracy of a rattlesnake's strike. Sight or heat sensing was sufficient on its own.

Other scientists have also needed to jam and unjam a sensory organ. For instance, Alejandro Purgue invented earmuffs for bullfrogs.

Now at Cornell University's Laboratory of Ornithology in Ithaca, Purgue had been studying the acoustics of North American bullfrogs. He began to suspect that the power of the male's commanding "ribit" came not from the throat but from vibrations in ear membranes. Purgue needed to damp and then release the membranes to document any differences in calls.

He tried smearing globs of standard laboratory silicon grease on the membranes to prevent them from vibrating. However, cleaning grease off the frogs' ears to reverse the effect turned out to be a nuisance. So, Purgue cut little ear pads out of the shock-absorbing foam used for inserts in running shoes. (He bought the foam at a supply company; no shoes were harmed in this experiment.) To link the pads, Purgue wound a tiny, tight coil of piano wire into a spring that would go over the frog's head.

This proved the sticking point of the design. "The angle between their ears was so shallow," says Purgue. "If your head was like that, you'd have trouble keeping your earmuffs on, too." He got a prototype to cover a bullfrog's head just long enough for him to admire the effect. "It just looked adorable," he says.

For the actual experiments, Purgue came up with a better solution: tidy devices that stay in place—the researcher's fingertips.

Purgue's hypothesis indeed proved correct. When he covered the frog's ears, the sound lost its power in midcroak.

SCIENTIFIC TRICKERY often demands a faux animal, and researchers turn puppeteers. One of the more famous of these stand-ins goes by the name RoboBadger. Its creator, Dan Blumstein of the University of California, Los Angeles, studied alarm calls of marmots and needed something harmless that the rodents would find alarming. Old taxidermy mounts have a long and distinguished history in the study of animal behavior, but the tricky part comes in revealing the stuffed menace to the creatures under study. It's not going to jog into place by itself, so researchers have an equally long and distinguished history of rigging drop cloths on strings to raise the curtain on the pretend threat.

Blumstein saw a way to render the arrival of his menace, a stuffed badger, a little more realistic. He dismantled a remote control model called Monster Truck and mounted the badger on the wheels and motor. RoboBadger now whirs into view on his own, though Blumstein has to select experimental sites near suitable trails that make good motorized-badger highways.

For a fake aerial predator, Blumstein developed Eagle Knievel, a customized, remote control model glider. "Marmots live in rocky places, and there's never a good place to land a model plane," he sighs. "I've spent days by now gluing him back together."

Lizard specialists like Diana K. Hews take a simpler approach. Based at Indiana State University in Terre Haute, she relies on one animal to provoke specific behaviors from its companions. She attaches a looped string to a fishing pole, lassos a real lizard, and then sets it down where she needs territorial conduct or flirtation. As she stands patiently nearby holding the end of the fishing pole, "they'll display, they'll even mate," Hews says.

Some experimenters fool their subjects with a fake environment rather than a phony or scientist-controlled animal. Paul Switzer of Eastern Illinois University in Charleston was investigating whether male dragonflies were more likely to switch to a new territory after a string of mating setbacks. He had to jinx a dragonfly's love life without changing its style.

"I was working at my aunt and uncle's farm, and they had plenty of sticks and wire," Switzer says. So, he set out sticks good for egg laying in dragonfly territories along a pond edge and attached them to a handle.

Switzer lurked nearby until a male started to impress a female prospect with a tour of a stick. Then Switzer dunked the stick out of sight. The male dragonflies seemed to search for it, but in the confusion, the females typically flew away. And yes, sexual discouragement did incline a male to seek new ground.

EVEN AFTER THE RESEARCH designer has worked out the central stage magic for an experiment, dozens of other challenges may loom ahead.

Aaron Krochmal, who grew up in New York City without a lot of personal wildlife experience, recalls that his surprises started when he received the first shipment of his test animals. For his dissertation, he'd devoted months to perfecting a protocol to test rattlesnakes' perception of environmental hiding spots, but he'd never had to manage a live rattler.

Disturbingly, his first shipment didn't arrive as he expected, with snakes in a bag, but with each coiled in a 2-pound deli container. Krochmal says, "My first thought was, 'How did they get them in there?' Which was followed rapidly by, 'How am I going to get them out?' "

He'd practiced basic handling routines on nonvenomous snakes, but not extraction methods from deli containers.

To anyone else who should suddenly confront this dilemma, Krochmal, now at Whitman College in Walla Walla, Wash., passes along his solution: Put the snakes in the fridge to chill them to sluggishness. Then, weigh down the top of a container—a length of rebar does nicely—while working loose the seal. With very long-handled clamps, move the container to the snake's new home, turn the container sideways, and gently squeeze.

Just keeping animals of unfamiliar species healthy in the laboratory can bring its own puzzles, says Hews. She recalls one husbandry crisis when she began rearing lots of little lizards in isolation containers. She fed them mutant fruit flies that can't fly, the cuisine that lizards usually enjoyed in her lab. In the solitary containers, though, many of the fruit flies evaded capture by roosting out of reach on the upper walls or outright escaping.

Hews therefore made her containers roost proof by dipping their upper sections in a nonstick coating that comes as a liquid.

Will Mackin of the University of North Carolina at Chapel Hill found a sweet way to measure how deep a bird called a shearwater dives. After finding the serious electronic wizardry that's available to be far too expensive, he came across journal articles describing equipment that costs about a dime. Mackin subsequently updated the technique. In his experiment, he sucked up confectioners' sugar to dust the insides of thin plastic tubes and then used a cigarette lighter to close one end. He attached the tube to a shearwater, and when the bird dove to forage, the increasing pressure drove water into the tube.

When Mackin recovered the tubes, he checked the ring marking the lowest point that water had dissolved the sugar and from that he could calculate the lowest point of the dive. He learned that the birds routinely dive about 7 meters.

Andrew Mason also figured out a modern version of old equipment as he monitored an *Ormia* fly's tracking the direction of a sound. Decades ago, German researchers who wanted to monitor the movement of an insect used a globe made for a streetlight. An optical sensor detected when an insect on top of the globe moved off center, and gearing underneath shifted the globe to keep the insect scrabbling on top. That system inspired Mason to paint irregular dots on a Ping-Pong ball and rest it in a modified cradle from an optical computer mouse. When he tethered the fly on top, the optical sensors picked up shifting dots and sent the information directly to a computer.

A BASIC QUESTION in studying social behavior is how to distinguish among the research animals. Even time-tested marking systems, such as colored bands for birds, can turn unexpectedly tricky. The commercial bands don't work on birds such as red-winged blackbirds, says Ken Yasukawa of Beloit College in Wisconsin. The birds fidget with the bands, squeezing them in their beaks until the bands eventually break off.

An idea from a colleague who custom-made bands for smaller birds, sent Yasukawa off to a craft store. He made his birds ankle bracelets, literally, from the plastic beads that melt together into colorful jewelry when pressed with an iron. He cuts a slit in the ring and pries it open just enough to slip it onto a bird's leg.

Yasukawa says a lot of his scientific equipment has come from Kmart, Ace Hardware, and craft stores, and that a good grasp of their offerings enhances creativity in scientific design. "Sometimes, I go just to look around," he says.

Other experimental behaviorists give close consideration to beauty products. Jill Mateo of the University of Chicago says that she's happy with the blue-black Lady Clairol for dyeing individual marks on the Belding's ground squirrels she studies.

Blumstein, however, has switched to a cattle dye for marking wild marmots. It's great, he says—easy to apply and long-lasting. Unfortunately, it dyes researchers' hands, too, a drawback when his field crews go into town for some sociability.

But that's not the most embarrassing social situation an innovative researcher has faced. Consider a string of experiments designed to see whether nonhuman primates have a sense of self. The researchers dyed a few pink streaks on an animal's ear or eyebrow and gave it a mirror. Apes appeared to notice something amiss, but experiments had not found similar reactions in monkeys. Marc Hauser of Harvard University wondered whether somehow changing the dye protocol, using more or different colors, might lead to different results.

He asked a student in his lab, whose hair on that day was orange, to recommend a nontoxic dye with lots of colors. "Manic Panic," she told him. The best local source turned out to be Hubba-Hubba, a store that advertises itself as "a boudoir of sin."

When Hauser bought four colors of dye, the clerk asked what he was going to do. He told her that they were for animals but immediately regretted the remark.

Back at the lab, Hauser found that he needed even more colors. When he arrived at Hubba-Hubba the next day, the clerk remembered him. "You were just in here yesterday," she said. "Now, what do you really want?"

To prepare budding young scientists for such eventualities, Steve Nowicki of Duke University in Durham, NC, starts raising the theme of experimental ingenuity in introductory biology courses. Plenty of concepts abound for great experiments, he tells his students, but they can lie around for decades stalled by some obstacle.

"The idea was there—what you had to do was make it work," he says.

MICHAEL BENSON

What Galileo *Saw*

FROM *THE NEW YORKER*

For the past eight years, the vintage spacecraft known as the *Galileo Orbiter* has been tracing a complex path between Jupiter's four large moons. During this time, it has made detailed scientific observations and taken thousands of high-resolution photographs, beaming them to Earth, half a billion miles away. On September 21st, 2003, *Galileo*'s extended tour of Jupiter's satellites will end, and it will hurtle directly toward the immense banded clouds and spinning storms of the largest planet in the solar system.

As the orbiter plummets toward Jupiter's atmosphere, several of its observational instruments will send a live transmission to Earth, and this data stream could prove highly illuminating. *Galileo* may be able to confirm the existence of a rocky ring close to the planet—a feature that has long been suspected. Other instruments will convey information about the density and composition of the mysterious, smokelike "gossamer rings" suspended inside the orbit of Amalthea, a moonlet near Jupiter.

At 2:57 PM Eastern Daylight Time, *Galileo* will be travelling at a speed of thirty miles per second, and its boxy octagonal frame will start glowing red. Seconds later, it will be white-hot. By 3 PM, many of its eighty-five thousand components will have separated from each other, and will continue to break up, becoming a hail of rapidly liquefying shrapnel. By the time the spacecraft's remains are three hundred miles inside Jupiter's atmosphere, where the temperature is twelve hundred degrees, all its aluminum components will have vaporized. At six hundred miles, its titanium parts will disintegrate. Jupiter is a gaseous planet, with a radius of forty-four thousand miles—big enough to contain all the other planets and moons of the solar system—and *Galileo* will have hardly penetrated its outermost atmospheric layer. Having just crossed Jupiter's threshold, it will vanish, leaving no clues of its earthly origin or its complicated mission.

Obliteration is precisely what NASA intends for the spacecraft. The reason is that *Galileo* may still harbor some signs of life on Earth: microorganisms that have survived since its launch from the Kennedy Space Center, in Florida, in 1989. If the orbiter were left to circle Jupiter after running out of propellant (barring an intervention, this would likely happen within a year), it might eventually crash into Europa, one of Jupiter's large moons. In 1996, *Galileo* conducted the first of eight close flybys of Europa, producing breathtaking pictures of its surface, which suggested that the moon has an immense ocean hidden beneath its frozen crust. These images have led to vociferous scientific debate about the prospects for life there; as a result, NASA officials decided that it was necessary to avoid the possibility of seeding Europa with alien life-forms. And so the craft has been programmed to commit suicide, guaranteeing a fiery, spectacular end to one of the most ambitious, tortured, and revelatory missions in the history of space exploration.

ALTHOUGH EUROPA wasn't the only target of *Galileo*'s camera during its years in space, its pictures of this weirdly fissured sphere—many of which show icebergs that apparently rafted into new positions before being refrozen into the moon's ice crust—produced euphoria among planetary scientists in the late nineties. They now speculate that Europa's global ocean may be more than thirty miles deep, which would mean that the moon has considerably more water than Earth. As Richard Terrile, a member of the NASA division that designed *Galileo*, has said, "How often is an ocean discovered? The last one was the Pacific, by Balboa, and that was five hundred years ago."

The orbiter also conducted forty flybys of planets and moons, far more than any other spacecraft. It was the first to swing close to an asteroid; the first to orbit one of the outer planets; the first to document fire fountains erupting from the surface of Jupiter's volcanic moon, Io; and the first to fly through a plume from Io, a lurid yellow-orange sphere with an estimated three hundred volcanoes erupting at any given time. In July, 1994, *Galileo* provided direct observation of fragments of the Shoemaker-Levy 9 comet slamming into

Jupiter; these collisions produced explosions more powerful than that of the largest H-bomb.

In recent years, when the mission was directed from Earth by a skeleton crew on a low budget and had absorbed more than four times as much of Jupiter's fierce radiation field as it had been designed to withstand, *Galileo*'s systems faltered frequently, but it continued to make discoveries. Last November, for example, its scanner registered the presence of up to nine tiny moons orbiting close to Jupiter. In June, 2000, it oddly failed to recognize the bright star cluster Delta Velorium, which flickers in Vela, a constellation that can be seen in the Southern Hemisphere. Subsequent observations from Earth confirmed that this group of five stars contains a dual-sun system, with one of its component parts periodically eclipsing the other, resulting in the variable light output that puzzled the spacecraft's instrument. *Galileo* thus became the first interplanetary space mission ever to make an interstellar discovery.

Conceived by NASA in the early seventies, *Galileo* had a rocky beginning; its early history was marked by a series of delays. Its entire flight plan had to be redesigned five times, both because its technical specifications kept changing and because the positions of the planets shifted between launch dates. It was trucked back and forth between California and Florida, and was disassembled, cleaned, stored, and then reassembled. Although the orbiter was an extremely sophisticated piece of technology for the seventies, when it finally went into space, in 1989, many of its systems were already out of date. (Its main processors were rebuilt versions of the RCA 1802 chip, which was used to run primitive video games like Pong.)

Galileo's most critical pre-launch problem was a woefully underpowered solid-fuel booster that could barely propel the craft out of Earth's orbit. It was able to get as far as Mars or Venus, but reaching the outer planets appeared to be impossible. *Galileo* had been specifically designed for shuttle deployment; after the explosion of the space shuttle *Challenger* in January, 1986, a newly safety-conscious NASA had decided that the orbiter's original, liquid-fuelled booster—which was more powerful but also potentially more dangerous than a solid-fuel device—couldn't be lofted alongside the shuttle's human cargo. The spacecraft seemed to be on the verge of a one-way trip to the Smithsonian.

Trajectory specialists at NASA's Jet Propulsion Laboratory set to work, attempting to figure out how to get *Galileo* to Jupiter with what amounted to a lawnmower engine under the hood. The man who eventually solved this puzzle was Roger Diehl.

I spoke with Diehl, who still works at the Jet Propulsion Laboratory, in July. He told me that his first idea was to get the spacecraft to Mars, and then use that planet's gravity to hurl it all the way to Jupiter. "I would go to bed at night, and my wife said she could even hear me talking about trajectories in my sleep," he recalled. But he eventually realized that because Mars had swung from its ideal position during one of *Galileo*'s launch delays, that approach wouldn't work. "It turns out that Mars is so small that if you go out of your way to fly by Mars to get a gravity assist you usually won't get a benefit," Diehl said. "So then I said, 'Well, let's launch to Venus.' "

This was hardly an obvious solution. Venus is in the inner solar system, and Jupiter is very far in the opposite direction. Moreover, this approach posed a significant thermal problem: *Galileo* had not been designed to travel closer to the sun before heading toward the frigid space around Jupiter. "If anyone had talked to a spacecraft person, there would have been a reluctance. They would have said, 'No, don't do that,' " Diehl said, laughing.

But he came up with a daring new flight plan anyway. Galileo would fly to Venus, curve back, swing around the Earth, then fly around Earth a *second* time exactly two years later; this trajectory would act like a slingshot, flinging *Galileo* all the way to Jupiter. Diehl realized that such a course would take several more years than the original plan, but he was undeterred. "I said to myself, 'I'm going to think of the problem as doing a tour of the planets of the solar system, with the goal of getting to Jupiter,' " he recalled. "I didn't care how many years it would take."

Diehl presented his boss, Bob Mitchell, with the unlikely scheme in August of 1986. Mitchell approved the concept, which was dubbed VEEGA, for "Venus Earth Earth Gravity Assist." Within days, Jet Propulsion Laboratory designers came up with a way to save *Galileo* from the harsh temperatures near Venus: they could attach lightweight, strategically placed sun shields that would protect it from intense heat.

In the next several months, two other scientists at the Jet Propulsion Laboratory, Lou D'Amario and Dennis Byrnes, substantially improved Diehl's initial concept; for example, they expanded *Galileo*'s itinerary, modifying its trajectory to make it fly past two asteroids. In the end, the VEEGA approach would require six years to propel the spacecraft to Jupiter, double the flight time of *Galileo*'s original plan.

Diehl considers his revision of *Galileo*'s trajectory, which effectively saved the mission, the highlight of his career. "My car license plate says 'VEEGA,' " Diehl said. "Every morning, I go out and I see the word."

GALILEO WAS SUCCESSFULLY deployed from the space shuttle *Atlantis* on October 18, 1989, seven years after its original launch date. It spent the next year making its detour to Venus. In December, 1990, *Galileo* began its "Earth-1" maneuver: the first Earth flyby. This happened to coincide with the buildup to the first Gulf War. NASA had to inform the North American Aerospace Defense Command that the blip that would appear on its radar screens on December 8th—an incredibly fast-moving object that might well seem to originate from the Middle East—was not an enemy missile but a robotic spacecraft coming from Venus.

Throughout the entire first part of its journey into space, *Galileo*'s umbrella-shaped high-gain antenna, intended to be its main communications link to the Earth from Jupiter, had remained snugly folded at one end of the craft. It was the largest such device ever to have been sent out of Earth's orbit. The plan was to deploy it only after the orbiter had receded far enough from the sun—because it, too, had originally been designed to operate in the bitter-cold temperatures of the outer solar system. In the meantime, the spacecraft would rely on a smaller, much slower antenna that was intended to be used only close to Earth.

In April, 1991, when *Galileo* was nearing the cooler climes of the asteroid belt, which is between Mars and Jupiter, the time had come to open the high-gain antenna and begin pulsing data toward Earth, at an optimal rate of a hundred and thirty-four kilobytes per second. *Galileo* was designed to have enough bandwidth to fire home one picture per minute, while also transmitting information from its other instruments.

But when the Jet Propulsion Laboratory finally ordered *Galileo* to open this key device, it stuck. Scientists running the mission were devastated: without a means of sending back high volumes of data, *Galileo* would be severely hobbled. Within a week of the antenna failure, two engineering teams were formed at the Jet Propulsion Laboratory. One was dedicated to getting the high-gain antenna unstuck. The other had to figure out how to rescue the mission without the use of the antenna; it was made up primarily of telecommunications specialists from the Deep Space Network. This division often provided NASA with a "million-mile screwdriver"—that is, a way of fixing a spacecraft by sending radio signals from Earth.

Leslie Deutsch, then the head of research and development for the Deep Space Network, is a garrulous but precise mathematician. "This was a crisis," he recalled in a recent conversation. "I got together with a few people, and we did some brainstorming. First, we said, 'Suppose we don't change anything. What's the data rate going to be when we get to Jupiter, if we have to use this low-gain antenna?' " The answer was ten bits per second, which translated to about one picture a month—and then only if *Galileo*'s ten other scientific instruments weren't in use. Such a data rate was pitifully inadequate; in space, complex phenomena must often be photographed many hundreds of times before they can be properly understood.

Instead of attempting to change the spacecraft's hardware, the Deep Space Network rescue squad began thinking about how it could improve *Galileo*'s information-processing capabilities. There was one possibility: *Galileo*'s fundamental software could be rewritten. To accomplish this feat, the onboard computer had to be powerful enough to handle the more advanced algorithms employed in the updated code. "The computer system on *Galileo* was ancient," Deutsch said. "So we looked into what kind of microprocessors were on board, and how much memory there was. And there was good news and bad news."

The bad news was that *Galileo*'s computer processors were so old that their original designers would need to be brought out of retirement for consultation. The good news was that, shortly before being launched into space, *Galileo* had been outfitted with twice as many memory chips as its designers originally intended; engineers had been worried that they were vulnerable to damage by radiation absorbed during the long journey in space. But after nineteen months in flight all the orbiter's memory chips were still functioning, which allowed Deutsch's team to do something that had never been attempted: change all a spacecraft's software applications in midflight. Updating the software would enable the team to introduce advanced data-compression techniques, which would help make it possible for *Galileo* to send useful pictures and other valuable information from Jupiter over the low-gain antenna. *Galileo* would now be capable of sending more than two hundred pictures per month, along with other data. This rate was considerably slower than originally planned, and some of *Galileo*'s objectives would have to be modified or abandoned. But the mission could still accomplish more than seventy per cent of its goals.

It took years, but by the time the orbiter completed its first sweep around Jupiter its software had been fully replaced. It was a move with unprecedented risks—"a complete brain transplant over a four-hundred-million-mile radio link," as one team paper put it—and any error could have meant losing the spacecraft. But the update was necessary, and the code transfer was flawless.

One problem remained: *Galileo* could collect information much faster than it could send it back. Its designers needed to find a way to store images, so that they could be slowly transmitted back to Earth. The orbiter, it turned out, had a tape recorder on board. Manufactured by the Odetics Corporation, in California, it was practically indistinguishable from the reel-to-reel recorders that were attached to the higher-end stereo systems of the sixties and seventies.

Though the machine was practically obsolete, it became one of *Galileo*'s most important features.

The recorder had been incorporated into the orbiter's design for one reason: to back up data from its atmospheric probe, which was scheduled to tunnel into Jupiter's clouds in 1995, when *Galileo* arrived at the planet's doorstep. This snub-nosed device would release its heat shield, deploy a parachute, and transmit information about Jupiter's atmosphere back to the orbiter as it sank into oblivion. The whole procedure was supposed to unfold over the course of an hour. During that time, the probe's findings would be relayed from *Galileo* back to Earth.

After the failure of the high-gain antenna, however, the tape recorder became a critical instrument. *Galileo*'s handlers at the Jet Propulsion Laboratory realized that it would be necessary to store all the incoming images and other scientific data gathered by its instruments during its flybys of Jupiter's moons. That information could then be fed into *Galileo*'s computers (using the new data-compression software) and slowly transmitted back to Earth during the months-long lulls when the craft was travelling between Jupiter's moons.

The magnetic tape spooled in *Galileo*'s tape recorder became a thread on which the mission's destiny hung. The entire system had been jerry-rigged, but it worked. *Galileo* began slowly transmitting spectacular images of Jupiter and its moons to Earth, where an upgraded antenna system picked up the spacecraft's slow, faint signals.

In March, 1996, the Jet Propulsion Laboratory team assigned the task of fixing the jammed antenna finally gave up. One analysis attributes the malfunction to a design flaw that was exacerbated by vibrations sustained when the antenna was hauled repeatedly between Florida and California during the years of launch delays.

EVEN BEFORE THE SCIENTISTS at the Jet Propulsion Laboratory finished updating *Galileo*'s software, the orbiter was not completely useless. In October, 1991, it took the first high-resolution images of an asteroid. Because the process of downloading photographs was so slow, it was instructed to send them back in fragments. Paul Geissler, a planetary geologist then at the University of Arizona's Lunar and Planetary Laboratory, was one of the few researchers allowed to view them as they came in, bit by bit. "It was wonderful—we were locked into a room and sworn to silence," Geissler said. "Because we didn't have the high-gain antenna, the data came in as what we call 'jailbars.' *Galileo* would send down a line, and then skip twenty lines, then send down another line, and then skip twenty lines and send down another line, and the issue was, is the asteroid in the frame at all, and should we use our precious bits to send down this frame or should we save it for the next frame?"

Geissler recalled the moment when his team realized that there was a tiny moon orbiting the asteroid. "In one of these jail-bars you could see Ida, and then it dropped off back into space again, and then there was another little blip. That's all we had. These particular jail-bars had three lines and then skipped a bunch, and this blip was in all three of the lines, so we were dead certain that it wasn't a cosmic-ray hit or anything like that. We knew there was something there. But we waited until another instrument on *Galileo* had a confirmation of it, and then we announced it."

Although astronomers had long believed that some asteroids have moon-lets, this was hard proof. It was also a reassuring illustration of what could still be achieved by *Galileo*, even in its extremely compromised state.

I asked Geissler, one of the leading image processors among planetary scientists, what it was like to see such unprecedented pictures before anyone else. He told me a story about the first complete shots of Ida, which had trickled in slowly, over a period of months in early 1994. "We had gotten two pictures of Ida up close, from different perspectives," he said. "As the spacecraft flew past the asteroid, it snapped a picture, at high resolution, and then it flew a little bit farther and then snapped another picture of the same region." Geissler realized that this separation allowed for the creation of a stereo image, which, when viewed properly, can give an object vivid three-dimensional form. "So I processed those pictures, and shot negatives of them, and brought them home—that was late on a Friday," he told me. "I had a darkroom at home, and later that night I made eight-by-tens of these two, and I had pinched a stereo-scope from work. I popped in these two wonderful eight-by-tens and became the first human being to see a stereo image of an asteroid at high resolution!" Geissler chuckled. "That entire weekend, anyone who came close to my door was dragged over—'Look at this!' You know, the mailman, the babysitter. That was really a thrill."

FOR DECADES, scientists have known that three of Jupiter's four large moons have high concentrations of frozen water. But only the hardiest optimists among them dared to speculate that liquid water could exist that far from the Sun. Europa's average surface temperature is estimated to be two hundred and sixty degrees below zero.

In 1979, the twin *Voyager* space probes flew past Jupiter at approximately ten times the speed of a rifle bullet. The closest they got to Europa was about a hundred thousand miles; *Galileo* has veered to within a hundred and twenty-four miles of the moon. Despite being so far away, the *Voyager* probes compiled a photographic record suggesting—indirectly—that Europa might be warmer below its icy surface. The most obvious clue was to be found on images of Io, Jupiter's innermost large moon. Firmly gripped by the tidal pull of its parent planet's gravity, yet yanked the other way by the shifting gravitational fields of two of its three sister moons, Io produces seemingly endless chains of active volcanoes. At three thousand degrees, they are far hotter at their source than any volcano on Earth. Io is the most volcanic object in the solar system; the mere proximity of such an excitable object to Europa suddenly rendered the idea of subsurface water more imaginable. If such active volcanism was present on Io, why couldn't there be similar eruptions on Europa's seabed?

The other *Voyager*-era clue was more subtle and mysterious. Long, looping chains of scalloped cracks snake across large spans of Europa's surface. These unusual patterns, which extend for hundreds of miles across the crystalline topography encircling the moon's poles, were already clearly visible in the *Voyager* images. In 1996, *Galileo* began taking highly detailed photographs of Europa. Scientists concluded with excitement that the fissures on Europa—which were dubbed "arcuate ridges"—were unique in the solar system.

Meanwhile, a handful of planetary geologists struggled to sort out what the curved lines on Europa's surface signified. One of them was Randy Tufts, a geologist at the University of Arizona. Tufts had been fascinated by those eerie ridges even before *Galileo* reached Jupiter. In a conversation I had with him a few years ago, he recalled that in the early nineties he had printed multiple copies of the low-resolution *Voyager* pictures and handed them out to his non-scientist friends—hoping that one of them might miraculously intuit the cause of the surface cracks. He had even taken the pictures to a glassblowing studio in downtown Tucson and asked the workers there if they had ever seen similar patterns. (They hadn't.) "I was just casting about for any kind of analogue," he told me.

In 1998, Tufts discovered an immense, gently curved fault line in the southern hemisphere of Europa. *Galileo* photographs revealed that the crack, which was subsequently named the Astypalaea Linea, extends about six hundred miles, which is comparable to the San Andreas Fault. This feature offered clear evidence that parts of Europa's crust were slowly moving—perhaps even floating.

That summer, it occurred to Tufts that the curvature exhibited by both the Astypalaea Linea and the arcuate ridges could be caused by the immense gravitational pull of Jupiter, which has three hundred times the mass of Earth. The linked curves of the arcuate ridges, he realized, could be explained by the fact that Jupiter does not exert a consistent amount of force on Europa. The planet pulls more strongly on the moon when the two bodies happen to be closer together. "Since Europa's elliptical orbit sometimes takes it farther away from Jupiter, the amount of stretching it undergoes kind of relaxes a little bit," he explained. Cracks start propagating—but then, as Europa recedes from Jupiter, they stop. Because Europa's Jupiter-facing hemisphere rocks back and forth during each orbit, by the time the gravitational stresses pick up again they're oriented in a slightly different direction.

With the help of Greg Hoppa, an orbital-dynamics specialist, Tufts plotted the effect of these fluctuating force levels; he ended up with looping cracks that look just like the ones on Europa. That was quite a breakthrough, but the team's next insight was even more significant: the whole process couldn't happen without the existence of a large body of subsurface water to exert tidal pressure from below. Ice crusted on solid rock could never be affected so much. The tides on Europa are much higher than those on Earth, reaching almost a hundred feet; when Jupiter pulls these enormous subsurface bulges of water in its direction, Tufts concluded, the ice on the surface begins to crack.

In the end, Tufts's insight is appealingly straightforward. By studying the elegant shapes on Europa's surface, he divined what lay beneath. "Later, I found myself sort of apologizing for its simplicity," he said. "And people said, 'Well, you know, some of the best ideas in science are very simple ones.' They're often so simple that everyone sees right through them." Tufts was excited by the idea that life might exist on Europa. "It always seemed to me that if we found life someplace else it would give us a vastly new perspective on existence," he told me. "And we would probably realize that we weren't quite so important as we thought we were." He frowned thoughtfully. "I mean, it might take us down a peg, which always could be useful."

Randy Tufts died last year, at the age of fifty-three, from a bone-marrow disorder. Not long before his death, he was working with scientists on plans for an orbiter that would investigate Europa's ocean more closely. In 2002, the project was cancelled, owing to budget cuts.

IN LATE JULY, I called Arthur C. Clarke at his home in Colombo, Sri Lanka, and asked him to comment on Galileo's impending death. Clarke has long been fascinated by Europa; it figures prominently in the sequels to *2001: A Space Odyssey*. In particular, I wondered if he shared NASA's concern that if

Galileo were to crash into the distant moon it could transfer microbes from Earth to Europa's ocean. Clarke didn't answer directly, instead suggesting that I read an old tale of his, "Before Eden," written in 1961. "It's all about the danger that we might contaminate new worlds," he said.

The story is about a scouting expedition to the South Pole of Venus, which is described as being "a hundred degrees hotter than Death Valley in midsummer." The expedition leaves behind a single human artifact—a bag of waste. It ends up contaminating a strange Venusian life-form that the expedition had discovered there, thus ending its evolution. I concluded that Clarke probably endorsed NASA's plan to destroy *Galileo*.

I spoke with Leslie Deutsch about his reaction to the decision, and he said he was initially angry, though he understood the rationale. Over the years, Deutsch admitted, he had become emotionally attached to the distant robot emissary, adding that this would be only the second time that NASA had deliberately destroyed a functioning spacecraft. When I asked Bill O'Neil, *Galileo*'s long-serving project director and one of the key architects of the effort to save the mission, what his reaction had been when he'd heard of the decision, he mulled it over for a few days, then sent me an e-mail. *Galileo*'s end would bring a personal sense of satisfaction at what had been achieved, he wrote. Still, he found it ironic that "Galileo Galilei only got house arrest by his sponsor the Roman Catholic Church for discovering things they didn't want to be true, whereas our Project Galileo gets a death sentence from NASA for its greatest discovery: the prospect of life on Europa."

BARBARA J. BECKER

Celestial Spectroscopy:
Making Reality Fit the Myth

FROM *SCIENCE*

In October 1859, German physicist Gustav Kirchhoff announced the results of his investigations with chemist Robert Bunsen on the dark lines that interrupt the otherwise continuous solar spectrum (1). These lines had puzzled practitioners and theorists alike since they were first observed in 1814 by German optician Josef von Fraunhofer (2).

Now it seemed that Bunsen and Kirchhoff had finally confirmed what others had long suspected, namely, that an individual metal produces its own characteristic pattern of bright spectral lines when it is burned. Furthermore, Kirchhoff asserted that Fraunhofer's lines "exist in consequence of the presence, in the incandescent atmosphere of the sun, of those substances which in the spectrum of a flame produce bright lines at the same place."

News of his claim spread quickly throughout the scientific world. In England, Bunsen's former student, Henry Enfield Roscoe, wrote to the secretary of the Royal Society, George Stokes (3): "Have you seen in the last no. of the *Annales* . . . a short note about Kirchoff's [*sic*] discovery . . . ?"

Soon, Roscoe was offering public lectures on the subject to interested scientists and laymen alike. After one such presentation to the Chemical Society, moderator Warren De La Rue remarked (4)

> [I]f we were to go to the sun, and to bring away some portions of it and ana-
> lyze them in our laboratories, we could not examine them more accurately
> than we can by this new mode of spectrum analysis. . . .

What really excited De La Rue, a stationer known for his photographs of the sun and moon, was the potential this method of analysis portended for astronomy. After all, French philosopher Auguste Comte in 1835 had clearly defined the domain of questions considered legitimate for Earth-bound observers to ask about the denizens of the celestial realm. "We can imagine the possibility of determining the shapes of stars, their distances, their sizes, and their movements," he declared, "but there is no means by which we will ever be able to examine their chemical composition"(5).

Despite this obvious hindrance, professional astronomers worked produc-tively and creatively throughout the 19th century to make many important dis-coveries: the successful determinations of solar and stellar parallax, the discovery of Neptune, and confirmation of the existence of an unseen com-panion to the star Sirius—to name a few. Meanwhile, their amateur colleagues, many of whom were no less serious about, or adept at, studying the sky, sifted patiently and tirelessly through the heavenly haystack hoping to be the first to find one of the proverbial needles that lay hidden there.

But De La Rue was right. Coupling the spectroscope to the astronomical telescope did revolutionize the way astronomy was performed, realigning the very boundaries of what astronomers considered to be acceptable research.

A quarter of a century after Kirchhoff's announcement, historian of astronomy Agnes Clerke marveled at the youthful audacity of a new science she called "astronomical or cosmical physics"(6). "It promises everything," she wrote, "it has already performed much; it will doubtless perform much more." And, she identified the enterprising English amateur astronomer William Huggins (1824–1910)—not Kirchhoff, Roscoe, or De La Rue—as one of stellar spectroscopy's principal founders.

A London silk merchant and self-taught amateur astronomer, Huggins joined the Royal Astronomical Society (RAS) in 1854. Soon after, he retired from commercial life and moved to the suburb of Tulse Hill, where he had a substantial observatory built to house his instruments. In 1856, as he recorded his first observations in a bound notebook, Huggins began his metamorphosis from curious dilettante to confident, self-directed observer. Along with his personal correspondence and publications, these notebooks help us trace the incremental career choices by which he, a recognized novice, shaped the inner dynamics of a new research agenda in a tradition-bound exact science.

Toward the end of his long career, Huggins waxed nostalgic as he witnessed his efforts being eclipsed by those of individuals like Hermann Carl Vogel (1841–1907), Edward Charles Pickering (1846–1919), and George Ellery Hale (1868–1938). With editorial assistance from his wife and collaborator, Margaret Lindsay Murray, he set to work putting his many achievements before the public, beginning in 1897 by publishing a personal retrospective entitled "The New Astronomy" in *Nineteenth Century*, a popular magazine of the day (7). In this stirring narrative—long a favorite source for Huggins's biographers—he recalls that soon after establishing his Tulse Hill observatory he became "a little dissatisfied with the routine character of ordinary astronomical work, and in a vague way sought about in my mind for the possibility of research upon the heavens in a new direction or by new methods." Luckily, he tells us, he had attended a lecture in 1862 given by his "friend and neighbour, Dr. W[illiam] Allen Miller."

Miller, a professor of chemistry at King's College, was repeating a well-received lecture on "The New Method of Spectrum Analysis" he had delivered some months earlier. In it, he drew particular attention to Fraunhofer's pioneering work on spectroscopy, which, Miller noted, had included an examination of the spectra of Venus and several prominent stars. But celestial spectroscopy was merely a prologue to the real purpose of Miller's talk, namely, to update the audience on what was currently known about spectrum analysis and how it was being used in the chemical laboratory.

Thirty-five years later, Huggins held the evening as his personal epiphany (7). The news of "Kirchhoff's great discovery," he proclaims, "was to me like the coming upon a spring of water in a dry and thirsty land . . ."

> A sudden impulse seized me, to suggest to [Miller] that we should return home together. On our way home I told him of what was in my mind, and asked him to join me in the attempt I was about to make, to apply Kirchhoff's methods to the stars.

According to Huggins, Miller obliged by coming to work in his observatory "on the first fine evening."

Unfortunately, Huggins's observatory notebooks contain no record of collaborative spectroscopic work with Miller in 1862. We know from correspondence and published sources that Miller was preoccupied at the time with his own research problems. Huggins, meanwhile, was taking advantage of an edge-on orientation of Saturn's rings to search for as-yet undiscovered moons of that planet. He did not find any—in fact, no one did—and the time and effort he expended on this, like many of his other short-lived or unsuccessful projects, eventually disappeared from the record.

Huggins and Miller soon communicated the results of their first efforts in celestial spectroscopy to the Royal Society, leaving no doubt they accomplished the work Huggins described in "The New Astronomy." But the vividness of his later recollections belies the fact that his early research program underwent not so much a radical transformation in 1862, as a gradual and fitful shift.

Operating on the periphery of the RAS, Huggins's independent research agenda was characterized by flexibility, alacrity, and openness rather than the traditional virtues of diligence, tenacity, and servitude which, in his view, restricted contemporary institution-bound observers. In the newly emerging field of celestial spectroscopy, no one, including Huggins, knew which line of research would prove the most fruitful. So, rather than, say, systematically catalog the spectra of all northern hemisphere stars, he chose to explore a number of different subjects in innovative and often technically challenging ways.

In August 1864, Huggins gamely tested the spectroscope's capacity to resolve questions about the nature of nebulae. It was a bold move that ultimately propelled him to a position of prestige and authority among fellow astronomers, capturing their imaginations and increasing their awareness of the potential of spectrum analysis to generate new knowledge about the heavens.

In May 1866, he became the first to analyze the spectrum of a nova (T Coronae). When the nova was lost from view, Huggins moved on to other projects: devising a method to observe solar prominences without an eclipse, spectroscopically determining the chemical composition of meteors, visually corroborating reported changes in lunar surface features, and even attaching a thermopile to his telescope to measure the heat of celestial bodies. In light of the mixed success of these projects, that he attempted them at all reveals much about his willingness, like a shrewd businessman, to take risks to establish himself securely in the forefront of research in the new astronomy.

One risky project that Huggins undertook in 1867 resulted in what is arguably one of the greatest, and certainly the most influential, of his many contributions to the new astronomy: his development of a spectroscopic method to determine the motion of a celestial body along its line of sight, a method integral to modern astronomical research. It is hard to conceive that in pursuing this project, Huggins relied only on direct visual comparisons of terrestrial and celestial spectra.

His notebook entries reveal this project as an audacious effort fraught with overwhelming mensurational and interpretative difficulties. But in announcing his findings to the Royal Society, Huggins expressed satisfaction that he had resolved the problems he had faced, stressed the care he had taken to insure the reliability of his measurements, and invoked the name of famed physicist James Clerk Maxwell to underscore the soundness of the theoretical foundation on which he based his conclusions. His tone was one of confidence and spirited adventure. Indeed, it could be said that Huggins's success in introducing this method to astronomical research lay in his ability to persuade his contemporaries that he had, in fact, accomplished what he claimed despite the overwhelming difficulties the method entailed.

Although few understood the physical theory on which his line-of-sight measurements were based, and implementing his method was largely beyond the resources and ability of many of his fellow amateurs; celestial mechanicians, including the Astronomer Royal, George Biddell Airy, recognized it as an aid to their mission of charting the positions and motions of celestial bodies.

It has been tempting for all who have read "The New Astronomy" to place it in a different category from Huggins's other published work: to see it as truer for its candor, more accurate for its detail, closer to the way things actually happened than any of his formal scientific papers. But, it is a synthetic account composed of selected events recalled some 35 years after the fact; like a well-laid string guiding us through a labyrinth after the puzzle has been solved, it excludes ambiguous choices, wrong turns, and dead ends.

Huggins was not a disinterested observer of the events that led to the growth of interest in analyzing the light of celestial bodies. He was an active participant in them. He could not tame his eclectic and opportunistic research style to fit the image of the methodical and systematic scientific investigator, but he could—and did—construct a more conforming public account of himself and his work, thus artfully situating his own benchmark contributions to celestial spectroscopy in the evolving lore of 19th-century spectrum analysis.

Huggins's unpublished notebooks and correspondence restore our contact with the visceral forces that drove him to venture outside the boundaries of acceptable research. They bring to light his instrumental and methodological ingenuity, his endless concern for priority, his power of persuasion, and his perseverance. Finally, they reveal him as a man who is more interesting and certainly more complex than the cautious, focused, and methodical individual portrayed in the public record.

1. G. Kirchhoff, *Philos. Mag.,* 4th ser., 21, 196 (1860).

2. Published in three segments. J. Fraunhofer, *Edinburgh J. Sci.* 7, 101–113 and 251–252 (1827); 8, 7–10 (1828).

3. H. E. Roscoe to G. G. Stokes, 24 February 1860, from *Stokes Papers,* Add. MS 7656.R788, Cambridge University Library.

4. W. De La Rue, *Chem. News* 4, 130 (1861).

5. A. Comte, *Cours de philosophie positive,* vol. 2 (Baillière, Paris, ed. 2, 1864), p. 6, translation by the author.

6. A. Clerke, *A Popular History of Astronomy During the Nineteenth Century* (Macmillan, New York, ed. 2, 1887), p. 180.

7. W. Huggins, *Nineteenth Century* 41, 907 (1897).

KEVIN PATTERSON

The Patient Predator

FROM *MOTHER JONES*

The creature arrived in the Arctic as it had spread itself around the world. It lay dormant in the lungs of someone apparently healthy enough to undertake such a journey, and then, when he was weakened, perhaps by hunger, or cold, or simple loneliness, it revived itself explosively. The sailor, or the missionary, or the trader found himself coughing paroxysmally and febrile, in some little iglu or tent, the bacterium streaming through his blood, to all his organs, and then a local Inuk had the misfortune to enter the shelter and leave. In Arviat, one of the Inuit communities that hug the shore of Hudson Bay, lives a woman I will call Therese Oopik, who carries the descendants of the creature within her. I visit Arviat in my work as an internist and she is my patient. We have been acquainted for a decade. Mostly we have had quiet conversations about contraception and bladder infections; these days we talk about this infection that seems likely to claim her.

She was beautiful when I first knew her, with a full-faced smile and glowing from the cold, but for two years now she has lost weight steadily and she looks as emaciated as a New York City fashion model. We joke about this. When I next visit there, she suggests, I might find a job for her.

Arviat is populated by 1,800 people, almost all Inuit, and it clings to the low rocks that stretch into the Arctic water like a collection of brightly painted aluminum-sided mollusks. The wind scours this place. Everyone knows everyone else and most of their problems. There have been episodic outbreaks of tuberculosis here, as in all the little communities in this part of the Canadian Arctic, ever since that first coughing sailor or missionary arrived. Tuberculosis has smoldered on notwithstanding the advent of antibiotics. The disease is primarily an expression of poverty and its consequences, especially overcrowding, and in the Arctic, these are usual. Latent infection endures in almost everyone older than 40. It revives itself regularly and seeps through the community; by the time a new outbreak is recognized there are usually dozens of new infections, some apparent, most already gone dormant.

Therese is 27 and does not know from whom she caught her illness. There are nearly as many possibilities as there are people around her. Half her left lung is taken up by a giant cavity full of the organism. Every time her sputum has been analyzed in the last two years, it has been found to be packed with the little rods of Mycobacterium tuberculosis that stain crimson when examined under the microscope. She coughs constantly and is so thin she is cold even inside, wrapped in blankets. After months of unsuccessful treatment in Arviat, she spent much of last winter in a hospital in southern Canada, but she continued to lose weight steadily. Her sputa remain resolutely positive. Finally, she insisted on going home, to her children, and to the tundra.

The bacteria inside the cavity in her lung are walled off by the surrounding scar tissue; blood does not flow there, and so antibiotics and the two arms of the immune system, white blood cells and antibodies, do not reach it. Therese's grandmother and many of the elders around her carry great scars across their chests from thoracoplasties, an operation to collapse such cavities. After World War II, with the development of effective antibiotics, chest surgeons were thrown out of work en masse (only to have their futures brightened anew by the rise in cigarette smoking and consequent lung cancers), and many experienced thoracic surgeons in the developed world today have never had to operate on a tuberculous lung. This is an astonishing and improbable state of affairs, really: Chest surgery was born of tuberculosis, and occupied itself almost entirely with the disease until 50 years ago. A chest surgeon then would not have believed TB could disappear for good, so quickly. He would have been right.

Our discussion about removing Therese's affected lung has been consequently tentative, but in any case she does not want an operation, or anything at all that involves more time away from her hamlet. She has the cachectic facies of a painting of a Victorian consumptive, Munch's Sick Child, perhaps. It is an uncomfortable feeling to find in her sickness the conventions of beauty—boniness and pallor. She is suspicious of doctors and nurses and takes her medication only episodically. People worry aloud about her children. This is a catastrophe in formation. A slow-moving freight train.

THIS CATASTROPHE threatens to erupt on a vastly larger scale than Arviat, to involve the entirety of North America. For the past 12 years I've treated patients in TB hot zones such as the Arctic, the Pacific Islands, inner-city America. What I and others on the front lines have to report is that TB is migrating out of these geographically remote or economically isolated communities and into the mainstream.

Perhaps this shouldn't surprise us. Tuberculosis infection has been so prevalent that for most of human history it was an almost normal, if often lethal, part of the human bio-niche. Hippocrates deemed it the most widespread disease of his time, and almost always fatal. In 1680, John Bunyan called TB "the captain of all these men of death" and indeed, until the development of antibiotics, 1 in 2 people who contracted it died. Between 1850 and 1950 it killed 1 billion people. Modigliani died of tuberculous meningitis, infection of the lining of the brain. Keats, a former medical student, remarked famously on the bright red and foreboding color of his sputum. Chopin, Chekov, Emily Brontë, Robert Louis Stevenson, Orwell, Kafka, Vivien Leigh: all claimed by TB. Then, for a mere 50 anomalous years, the disease ebbed and a lingering cough was no longer cause for immediate terror.

Once antibiotics were mass-produced in the late 1940s, First World transmission rates fell steadily; when latent infections reemerged, they were usually caught and cured. Such hidden reservoirs of TB dwindled as those infected either aged and died of other causes, or were identified through positive skin tests and treated. By the early 1970s, the problem was seen as past, and U.S. public-health authorities charged with monitoring and treating residual outbreaks began losing their funding. Surveillance for TB became more lax, supervision of therapy more haphazard. Such slipshod treatment has led not only to new infections but, more terrifyingly, to the emergence of multidrug-resistant TB strains, or MDR-TB, which a recent Harvard study called "Ebola with wings," and is now found in the United States and more than 100 other countries.

Several other events have acted to reignite the old problem known as phthisis, consumption, or the white plague (named for the pallor it causes). Just as public-health spending on TB was declining, immigration from areas where the disease has never been much beaten back—from Asia, Africa, and especially from Russia—steadily rose. Despite a decade-long effort mounted by the

World Health Organization, the global rate of TB continues to climb about 3 percent a year. Currently, then, 8 million people worldwide fall ill with the disease each year and, because only a quarter receive effective treatment, 3 million die. Furthermore, a third of the world's population—another 2 billion people—carry latent infection, each of whom has about a 10 percent chance of it activating in their lifetime. When TB does activate, it does so insidiously, often simply as a persistent but infectious cough, and on average, an active case will infect 20 other people before it is identified and treated.

Unless Americans are willing to adopt suffocatingly draconian immigration policies, the likelihood is that with globalization TB will again become epidemic here, in the same way that HIV moved from Africa to take root throughout the world. Suffering does not localize. When we engage with the world, we engage, inescapably and absolutely, with the world's infections. And the most devastating infection in the world is not Ebola or Lyme disease, West Nile virus or even HIV, but tuberculosis.

MYCOBACTERIUM TUBERCULOSIS is descended from similar organisms that have inhabited cattle for eons. About 8,000 years ago the organism spread to humans. It has since adapted to us, tweaking its genetic arrangement to optimize its capacity to spread and persist.

Its virulence is largely a consequence of the thick, waxy cell wall that surrounds the organism like a rind, making it substantially resistant to the immune system. When the organism is inhaled, it lodges in the alveoli of the lungs, the little air sacs that exchange oxygen for carbon dioxide within the blood. When the immune system identifies the presence of this foreign protein, it recruits white blood cells called macrophages ("big eaters" from the Greek) to engulf and bathe the invader with toxic chemicals. Despite the macrophages' defenses, the creature, because of its thick rind, often survives and slowly replicates itself until each macrophage is so full of tuberculosis bacteria that the cell bursts and dies. The organisms drift on, to be engulfed in turn. Most of the time an equilibrium is attained, nearly as many tuberculosis bacteria are killed as survive, and the spread of the creature is checked. This is the latent phase, and it can go on for decades until for some reason, or no obvious reason at all, the balance is tipped.

The balance might tip because the immune system is weakened by stress, or by pregnancy, or by another infection (especially HIV), or by starvation, or simply by age. In any case, when it does, the bacteria overwhelm the immune system. They begin to fill the adjacent alveoli and block the exchange of oxygen and carbon dioxide. The patient becomes short of breath. The immune system's efforts to fight off the infection inflame the walls of the lung, which come to bleed easily. The patient develops a fever, and she notices streaks of blood in her sputum. Initially, as Keats knew, this blood will be dark red and venous; later, as infection erodes its way into arteries, it becomes bright crimson. If the affected artery is a large one, this may be what is called, chillingly, a "pre-terminal" event. Katherine Mansfield expired in a great geyser this way.

Even if the artery isn't large, now the seed is in the wind and the creature in the blood—it swims to every organ system: the brain, the kidneys, the intestines, the bones, the ovaries, the skin, the uterus, the heart. Any medical textbook more than a generation or two old lists tuberculosis as a possible cause of almost any organ dysfunction, and usually high up on that list.

ONE OF THE GREAT CALAMITIES for all mankind, tuberculosis was simply the worst thing that ever happened to the Inuit, and to indigenous Americans more generally. Nobody knows why it has been so prevalent among them. There are many possible explanations: the small, crowded snow and skin dwellings they lived in; the lung diseases caused by the smoky seal-blubber lamps they lit the iglus with; a genetic vulnerability; and, most persuasively, the chronic, recurrent famine that always lurked on the edge of the ice floes. The Inuit called it puvaluq, or bad lung, and once introduced by Europeans it infected most of the Inuit in the Arctic. Across the continent it killed many thousands—perhaps millions—more Native Americans, largely clearing the Great Plains; when settlers spread across them in the mid-1800s, they often commented on the wide-open and fertile grasslands, uninterrupted by signs of human habitation. But for the creature, the land would have been as it had been only a few years earlier—alive with children and hunters and women growing corn.

The small desperate bands of remaining Plains Indians were herded into crowded reservations that accelerated the contagion process already well under way. In Kivalliq, where I work, the Inuit were compelled by the Canadian government in the early '50s to move into the dismal little settlements it had erected on the shore of Hudson Bay, largely to facilitate attempts to control the infection. Throughout the '50s and '60s, ships with mobile X-ray units visited every summer, and those found to have active disease were evacuated to sanatoriums in the south. Children later returned to their barely remembered parents were often unable even to speak to them, their language lost in the land of drive-in movies. By the late 1960s, the last of the nomadic hunting families had come in off the ice and settled in the clapboard houses provided them by the government, there to begin the cycle of acculturation that continues today.

In the hamlets of Coral Harbour and Arviat, there are now active epidemics. Five years ago it was in Repulse Bay, a few years before that, Chesterfield Inlet. The Inuit's historic exposure rate to TB was so high that it can be expected to recur reliably for decades to come. TB remains endemic among most Native Americans, with an annual incidence that in 1987, for example, was 400 times higher than among urban Caucasians. Not that resources haven't been thrown at the problem—in Kivalliq the annual spending per capita on health approaches $15,000.

But tuberculosis isn't like other communicable diseases that can be cured with a few doses of inexpensive and safe medicine. Like a storm forming from the coalescence of several weather systems, TB is the perfect pathogen in many respects: It has long latent asymptomatic phases, allowing it to dwell on in populations; it is spread invisibly, in the air, so infection-control measures like hand washing have no effect on transmission; and when it takes hold it erodes every single organ system. And its perfection is preserved most of all by its ability to shrug off our medicines within a few decades of their development.

POOLS OF PREVALENT tuberculosis infection surround North America from every side, including the islands of the Pacific Ocean, the American possessions of Saipan, Guam, and Samoa among them. The problems of poverty and crowding among the Inuit are echoed on these islands; the only real difference is the temperature, but this ceases to matter to the infected—as the disease progresses, one feels as if on fire wherever one is.

A few winters ago I spent three months working in Saipan, the largest of the Northern Mariana Islands, which rises out of the deepest water on the planet as a long and verdant mountainous ridge. Most people in Saipan live close to the sea, descendants of ancient mariners who undertook thousand-mile-long open-water passages in outrigger canoes, without compasses or maps, as we understand them. A kind of poverty exists there today that did not when the people were less wealthy. The island is more crowded than it ever has been, and rests uneasily between an Asian and an American sensibility. It has the diseases of both existences and the cultural stability of neither—the perfect picture of globalization.

The Marianas are remnants of the Pacific empire that America won from Japan and are now a commonwealth of the United States, like Puerto Rico; those born here are American citizens. Most of the people who live in Saipan are not, however, but rather come here as contract workers for the garment industries on the island, which value the combination of "Made in the USA" labels and cheap labor. The workers migrate north from the Marshall Islands, and west from China, Thailand, and Bangladesh.

Saipan's TB clinic processes a steady stream of arrivals found to have shadows on their mandatory screening chest X-ray. Legal workers are motivated to comply with their therapy—if not, they are sent home expeditiously. But illegal workers represent a thornier ethical and political problem—as they do throughout the developed world—and one of the principal obstacles to controlling TB. For wherever there is suspicion between public-health authorities (the term itself is illuminating) and the infected, TB thrives. Illegal workers, migrants, intravenous drug users, all of whom dwell in the various demimondes of a fractured society, are among the most likely to be infected by TB. And dwellers of any demimonde recoil instinctively from a stranger's knock on the door. This is a survival skill.

Among Saipan's indigenous Chamorro—much wealthier and U.S. citizens—TB also endures and flares periodically, not yielding much over time to the efforts of Artin Mahmoudi, who ran the clinic. Dr. Mahmoudi examined the DNA from TB strains he obtained from the Chamorro and the migrant workers. Despite the widely held prejudice against the migrants, he found that, for the most part, the TB found among the Chamorro is not acquired from them, but from latent infections that have activated. There is enormous potential here for new strains of TB to burst out of the peeling-paint dormitories that house the migrants. But this has not occurred. Yet. In the meantime, Mahmoudi holds his breath. And wrestles with that dragon, not yet loose.

Mahmoudi is Baha'i, originally from Iran; three family members were killed following the overthrow of the Shah. He is a handsome and graciously mannered man; his personality dominated the hospital on Saipan. This comes across as high-handedness to some. An acquaintance once told me, before I'd met Mahmoudi, "You do have to spend a fair amount of time telling him how smart he is." It was clear to me, however, that he elevated the standard of Saipan's medical care far beyond what it otherwise would have been.

His desk, in the office he shared with three other internists, all crowded together just like in a college-dorm study center, was littered with books on the history of TB and CDC reports on new resistant strains. His faith makes him suited for his task; he is part zealot, part missionary. His is the sort of effort that could change the picture of transmission worldwide, but there are no other Mahmoudis anywhere close to here. Elsewhere, in the Marshalls, for instance,

TB is simply an observed and not much discussed fact. People cough, they sweat, they die.

Such indifference is a measure of the disease's intractability. The first effective anti-TB drug, streptomycin, came into use in 1948. By 1950, resistant strains were commonly observed. Other antibiotics were developed in rapid succession, but TB proved fully capable of becoming resistant to any of these agents, and quickly. A strategy was devised to bombard a patient with four drugs simultaneously and then to continue therapy for six months with at least two drugs that the patient shows no resistance to. All such drugs can have debilitating side effects, including color blindness, numbness, deafness. It's easy to imagine why an Inuit hunter would eschew treatment, why anyone would. In practice, people take the medication reliably as long as they feel unwell, and when they feel better they stop.

Ideally, then, the drugs are administered by a health professional, and their ingestion witnessed. When this approach—called Directly Observed Therapy, Short Course, or DOTS—is used, resistance is dramatically lessened. But when treatment is sporadic, when DOTS therapy isn't pursued until completion—and this is common, even in the United States—resistance becomes inevitable.

The facility with which Mycobacterium tuberculosis develops resistance surpasses that of nearly any other organism; it is one of its defining characteristics. When TB becomes resistant to multiple drugs, the cure rate, even under the best circumstances, falls from 98 percent to 60 to 70 percent. In the Third World, treatment of MDR-TB is so expensive—up to $10,000 per person—and difficult to deliver that the World Health Organization recommends that antibiotics should be dispensed only to patients whose bacteria is still drug sensitive, and then only in the context of a DOTS program. Unfortunately, DOTS is not practiced in a third of the world's nations.

Mahmoudi introduced the DOTS strategy to Saipan, and with it at least those identified are reliably treated. But the countervailing effects of increased immigration and the rise of chronic diseases among the Chamorro have served to check dramatic progress. The irony in the Pacific Islands is that, though still beset by traditional tropical infections such as malaria, dengue fever, and elephantiasis, the new great killer is the combination of obesity, diabetes, and vascular disease. These diseases of affluence have not so much supplanted the old diseases of deprivation as teamed up with them. Diabetes—epidemic among Pacific Islanders, as it is becoming among the other great traditional maritime people, the Inuit—or any malady that causes immunosuppression, makes TB, aggressive enough in the healthy, vastly more threatening. It is in keeping with its character that TB would find a way in the modernizing Pacific to exploit both the enduring poverty of the area and the consequence of imported and too plentiful food.

I spent my evenings walking along Beach Road, down by the American and Japanese tourist hotels. Just off Beach Road are brothels and massage parlors. The women who work here are garrulous. Most were garment workers but grew tired of minimum wage and breathing cotton fibers all day long. Astonishingly, there is very little HIV here—so far. Everyone knows that is going to change. Mahmoudi just shook his head when he discussed the implications HIV will bring to the problem of TB here.

When HIV emerged 20 years ago, it helped cause the incidence of TB in the continental United States to rise 20 percent among adults and 35 percent among children. But things are far more dire in the developing world: In Africa, for example, tuberculosis is the disease that kills the vast preponderance of AIDS patients. HIV and tuberculosis are almost co-infections—each renders its victims more susceptible to the other. From the moment a person becomes infected with HIV, he is 800 times more likely to suffer activation of latent TB. This is because the type of immunity that AIDS specifically assaults, called the cell-mediated immune system, is precisely the arm that normally resists TB infection. At the same time, TB imposes a certain type of immunodeficiency on those it infects that renders them far more susceptible to HIV. The twin catastrophes of HIV and TB share far more than geography; they are as intimately complicit in each other's evil as Leopold and Loeb.

IN AUGUST, 2002, I traveled from my home on Saltspring Island, British Columbia, to New York on a Cathay Pacific flight that originated in Hong Kong, connecting through Vancouver. The passengers around me were exhausted from the 14 hours they had already spent flying, and the sound of rasping snores filled the cabin, notwithstanding *The Royal Tenenbaums* and their efforts to distract us. In the orange high-altitude morning light, Gene Hackman barked silently on the bulkhead screen in front of us. Behind me a child coughed steadily, suffering, no doubt, from an innocent viral respiratory-tract infection, or perhaps some sinusitis.

The passengers around me were distinguishable from the men and women I treated in Saipan only by the authenticity of their Rolexes. They appeared to have prospered in China's economic boom, probably now resident in Hong Kong or Singapore, but the adults were likely born into conditions similar to those of the Saipan garment workers. How many degrees of separation from TB could I possibly hope to be?

I had come to the New York area to see Lee Reichman, who knows and admires Artin Mahmoudi. A former president of the American Lung Association, Reichman is now the executive director of the New Jersey Medical School's National Tuberculosis Center, where he oversees 78 physicians and nine field-workers. He is 64, has been working with TB for the last 30 years, and is so deeply alarmed that in the fall of 2001 he published *Timebomb: The Global Epidemic of Multi-Drug-Resistant Tuberculosis* to draw attention to the problem.

One of the reasons that Reichman is held in such high regard is that in the 1970s he was virtually alone in warning American public-health authorities of the disaster that would come if they slashed TB funding. Sure enough, in 1986, thousands of new cases of active TB, many MDR-TB, appeared in U.S. cities; in 10 years the rate of new infection increased by 20 percent. Not only was the rate of TB infection increasing, but it was appearing in odd forms—outside the lungs, as meningitis, or in the kidneys, for instance—more frequently than normal. The cause was HIV. Checking this surge cost more than a billion dollars.

But once again, Reichman sees what he calls "the U-shaped curve of concern." Thinking that they've successfully combated the latest TB outbreak, the medical profession is again treating the illness haphazardly. "We know this from the development of MDR-TB," he says. "Every case of MDR-TB is evidence of inept medical therapy; these strains do not occur naturally." A study Mahmoudi published found that in 1993 doctors in a Denver hospital made an average of 3.9 treatment errors per TB patient; this poor treatment of an outbreak that began among migrant workers gave rise to a 75 percent increase in Colorado cases—including some that are MDR—in the past three years. "Sadly," Reichman notes, "little has changed. In the states of Virginia and Washington, TB medications are still not provided free to patients. This represents a violation of every principle of TB care. You have to make it as likely as possible that the patient will take the medicine for the necessary length of time. And the people with the least money are the most likely to have TB."

We sit in his office in central Newark. He is animated and arch, not as cool as Mahmoudi, more accessible. He is dismayed by the indifference of the public to its own peril, but it is the acquiescent dismay of an older man. He shrugs. There are other problems just as large, right now. He wants to tell me about the Russian prisons.

In Russia there are 1 million men and women in stygian cages, coughing into the close and crowded air. Reichman, who visited the prisons in 1998, thinks that essentially all of them have either active or latent TB by the time they are released. He was flabbergasted by the experience. The treatment that Russian prisons employ is more of a gesture than therapy, and it is a disastrous gesture: occasional dosing with inadequate antibiotics, giving rise to organisms with resistance so broad as to be almost untreatable. It is only one of the horrors of Russian prisons, but it is the one that threatens the world outside those concrete walls.

Every year 300,000 Russian prisoners are released. Then they, and the creature within them, move away, understandably, from those cages, often with some velocity. Inevitably, some are propelled across the Atlantic. Barry Kreiswirth is a microbiologist who works with Reichman. His preoccupation is collecting and categorizing the DNA fingerprints of different strains of TB; he has 16,000 now, the world's largest collection, he told me with great pride.

In Tomsk, Siberia, is a prison rife with TB, including a strain that arose there called W148 that is almost always resistant to all commonly used anti-TB antibiotics. In March 2000 Kreiswirth was asked to examine a new strain of resistant TB found in New York. It was W148. He was stunned. He contacted the New York Public Health Department. The strain was found in a Russian immigrant. W148 was here. Suffering does not localize.

This example is just one among thousands. The World Health Organization estimates that 50 percent of the world's refugees are infected with TB. And at least half of all diagnosed TB cases are found among foreign-born people who have moved to industrialized countries. Legal immigrants to America over the age of 15 must have a chest X-ray upon entry. But this does not catch those for whom the disease is not developed enough to show up on film, or adopted children who come with active TB, or business travelers, or tourists, or illegal aliens. In the 1990s, cases among foreign-born Americans rose from 29 percent to 41.6 percent. So it is not surprising that U.S. infections are, thus far, concentrated in areas favored by new arrivals. Strains of TB once found only in Mexico have migrated to Texas. In 2000, some 16,000 people living in the United States developed active TB; half were foreign born, with 1,332—or 8.3 percent—living in New York City. In 2001, 36 percent of all reported MDR cases were in New York City and California.

But just as it spread from Siberia and Senegal, TB will not stay contained within pockets of America for long. For now it is chiefly found among the marginal and the overcrowded, epidemic, for example, among the homeless and in our prison system, where 2 million Americans reside and 600,000 more work.

But prisoners are paroled and guards go home. In 1993, Reichman writes, "one New York county found that 24 percent of identified cases occurred in jail inmates, former inmates, jail employees, or community contacts."

Fear of MDR-TB has led to a different kind of incarceration. In the last few years U.S. health authorities have committed hundreds of patients to hospitals to ensure that they complete their course of medicine. The notion of treating sane patients against their will is deeply discomfiting to most physicians. The laws that allow for this date from the Victorian era, and the idea seems similarly obsolete. But the fact that this measure continues to be resorted to underscores the threat that MDR-TB represents. There hasn't been a new class of anti-TB antibiotics developed since the late 1960s; new drugs are estimated to be at least seven years away. Fifty million people have MDR-TB, 2.5 percent of all TB cases. But in certain countries, 20 percent are MDR. And in some places, including in New York City, strains exist that are resistant to all known antibiotics.

Reichman is blunt: "We sit on the edge of a potential catastrophe. Government doesn't take this problem seriously, physicians do not reliably treat this problem, and the public thinks TB isn't sexy enough to merit its attention. Ebola kills a couple of hundred people and there are movies about it. TB kills 2 million people a year already, more than HIV and malaria, more than any other single infection. And it is curable and has been curable since 1948. This inattention is just what this infection needs to explode, for the time bomb to go off!"

To the extent that anyone is working to forestall such an explosion, it is people like Rebecca Stevens. She is a field-worker for Reichman, administering DOTS, and I went with her as she fanned out into central Newark, policing the compliance of 28 patients. There is some irony to this: Both Rebecca and I were previously diagnosed with latent tuberculosis, a troublesomely common affliction among health care workers. We both hope that ours are drug-sensitive organisms. Treatment among latent patients is always speculative as it is impossible to test them for drug sensitivity. It is conceivable that one day either Rebecca or I might find ourselves with a Rebecca in our own lives, dropping by to watch us take our pills.

We visit the apartment of sisters Chariesse and Mona Greene, and the nine children they care for between them. The kids range from five months to 10 years; six have developed TB. Chariesse is sick, and weighs 94 pounds, even less than Therese Oopik. Out of the hospital for a week, she is tired and worried about her infant niece, who is still hospitalized with tuberculous meningitis. Rebecca says the child does not look well.

Mona had tuberculous meningitis when she was 12, she tells me. Chariesse may have caught TB from Mona at that time, or even years earlier. It is not possible to know. The children probably caught TB from Chariesse. Another boy Mona babysits is in the hospital, too. He's doing better than her niece. I tell her I'm glad. She nods. She hollers at one of her daughters for not picking up her clothes. Rebecca yells back that the hollering is giving her a headache. The house quiets as Rebecca mixes the drugs in with jam for the younger children.

The children ask Rebecca for their cans of Boost, a dietary supplement she brings them when she visits. "Mona is what keeps this situation together," Rebecca tells me privately. "Viewed from their own situation, they're doing okay. From my situation, or yours, maybe not, but from theirs, they're all right."

The apartment has a manic, agitated quality to it; the children are both excited by visitors and shy. Chariesse and Mona are in their 20s, and seem embarrassed about the chaos around them. Both have had drug problems on and off for a long time. Currently not, they say, and Rebecca is inclined to believe them. Chariesse is too sick, and since she got sick, everything pretty much falls to Mona. Things are going too well, Rebecca says, for Mona to be using much.

Too well is not that well, really. They are being evicted. The $800 in welfare they receive is not enough to pay for both food and the apartment they live in, where rats leap from the cupboards when doors are opened too quickly. Mona and Rebecca have a long conversation about the homeless shelters in the area. There are boxes stacked along the living room wall, ready to move their few possessions, and clothes and mattresses on the floor. As we prepare to leave, to distribute the antituberculosis medications to other patients, Mona says she doesn't know how they are going to do it, where they will live.

The last time I saw Therese, she told me she tried not to think about the future, about how all this would end. She enjoyed being home and tasting the Arctic char. The snow would melt soon, and there would be walrus and beluga whale to eat. She knew the peace wouldn't last. And it won't.

MICHAEL POLLAN

Cruising on the Ark of Taste

FROM *MOTHER JONES*

The first time I heard about the Slow Food movement, recently arrived on our shores from its native Italy, I thought the whole idea sounded cute. Here were a bunch of well-heeled foodies getting together to celebrate the fast-disappearing virtues of the slow life: traditional foods traditionally prepared and eaten at leisurely communal meals. They aimed to save endangered domestic plants and animals—the Vesuvian apricot, the Piedmontese cow—by eating them. Slow Foodies were antiquarian connoisseurs, I figured, with about as much to contribute to the debate over the food system as a colloquium of buggy whip fanciers might have to add to the debate over SUVs.

Certainly it's hard to take seriously a political movement that has a snail for a mascot and a manifesto calling for "a firm defense of quiet material pleasure." But after learning more about it recently, I've come to think that Slow Food might actually have a serious contribution to make to the debate over environmentalism and globalism. Not that any self-respecting member of Slow Food would ever want you to think they take themselves that seriously; pleasure is at the very heart of their movement, which is dedicated to the proposition that the best way to defend the planet's cultural and biological diversity is to enjoy it at the table, slowly. Whether it means to or not, Slow Food is mounting a provocative challenge to some stale lefty assumptions about consumption, free trade, and the place (if any) of pleasure in our politics.

As its name suggests, Slow Food is a reactionary organization, but reactionary in the best sense. It took shape in 1986 in the brain of Carlo Petrini, a leftwing Italian journalist dismayed by the opening of a McDonald's on the Piazza di Spagna in Rome—and perhaps equally dismayed by the hangdog dourness of his comrades on the left. After years of activism he had come to the conclusion that "those who suffer for others do more damage to humanity than those who enjoy themselves," as he recently told a group of journalists. "Pleasure is a way of being at one with yourself and others." So rather than picket McDonald's' new outpost in the heart of Rome, or drive a tractor through it à la José Bové, Petrini organized a group of like-minded activist-cum-sybarites to simply celebrate all those qualities that McDonald's' inexorable drive toward the homogenization of world taste threatens: the staunchly local, the irreplaceably unique, the leisurely and communal. His (so-very-Italian) idea was to launch a political movement conceived under the signs of pleasure and irony: Dionysus meets Dario Fo.

In 2003, McDonald's is still serving Happy Meals by the Spanish Steps (though Petrini did persuade the company to hold the golden arches), yet Slow Food has emerged as a thriving international organization, with more than 65,000 members in 45 countries, a successful publishing operation (Slow Food's *Gambero Rosso*—an indispensable Zagat-like guide to Italian food and wine—pays most of the bills), and the Salone del Gusto, a biannual trade show that brought 126,000 eaters together with artisanal food producers in Turin last October. Just as important, Slow Food has launched a handful of decidedly eccentric institutions and ideas—the Ark of Taste, the presidia, "eco-gastronomy," and "virtuous globalization." Unpack these terms and you have a pretty good idea what's afoot—and at stake.

The Ark of Taste is basically the list of endangered food plants and animals that Slow Food has resolved to defend against the rising global tide of McDonald's-ization. Some American passengers recently added to the Ark include Iroquois white corn, the red abalone, the Narragansett turkey, the Sun Crest peach, and the Delaware Bay oyster. We've come to think of biodiversity as a biological crisis of wild species, but the survival of the domesticated species we've depended on for centuries is no less important. For one thing, when the latest patented hybrid-corn variety meets its bacterial or fungal match, as all monocultures sooner or later do, breeders will need these heirloom varieties to refresh the gene pool. Should that Iroquois white corn fall out of production, as it very nearly did a decade ago, an irreplaceable and quite possibly crucial set of corn genes would be lost to the world.

Of course seed-saver groups have been around for a while now, preserving heirloom varieties from the onslaught of patented hybrids, but Slow Food takes that project a step further. The movement understands that every set of genes on its Ark of Taste encodes not only a set of biological traits but a set of cultural practices as well, and in some cases even a way of life. Take the example of Iroquois white corn. By working to find new markets for this ancient culti- var, Slow Food (along with the Collective Heritage Institute, its partner in this particular project) is ensuring the livelihood of the Native Americans who grow, roast, and grind this corn (on the Cattaraugus reservation in western New York) and the specific culinary and spiritual uses that corn has been selected over hundreds of years to support. "Save the Corsican Chèvre!" might not sound like a life-and-death battle cry, until you realize, as Slow Food teaches, that as those goats go, so goes something greater: a specific, irreplace- able mode that a particular people have devised for living on, and off, a partic- ular corner of the earth. Save the genes, and you help save the land and the culture as well.

Slow Food recognizes that the best place to preserve biological and cultural diversity is not in museums or zoos but, as it were, on our plates: by finding new markets for precious-but-obscure foodstuffs. This is what is meant by "eco-gastronomy." Slow Food features the foods and their producers at its Salone del Gusto (Hall of Taste), and organizes tastings at its local chapters (called Convivia), where an effort is made to educate palates in the course of exercising them at a feast. This emphasis on celebration and connoisseurship has left Slow Food open to charges of elitism, but the organization has worked hard to reach beyond the affluent foodie crowd. Slow Food USA has launched a garden project for public schools, and a great many of the foods it has champi- oned in the United States are distinctly populist and often cheap: Barbecue and beer are as much a part of the movement as endangered oysters and rare sakes. "To me, Slow Food is spending a few quarters on a Spitzenberg apple instead of a Red Delicious," says Patrick Martins, the energetic young director of Slow Food USA. "It doesn't have to be an everyday thing." Sure, fast food is always going to be "cheaper" than slow food, but only because the real costs of the industrial food chain—to the health of the environment, the consumer, and the worker—never get counted.

Even Slow Food's concern with connoisseurship is not as effete as it might sound. Along with the industrialization of our food system has come an industrialization of eating, and the former won't be effectively countered until people have rejected the latter. Slow Food aims to teach us to taste what makes Iroquois corn special (it's wonderful stuff, with an earthy, sweet, extra- corny flavor that makes commercial corn products taste pallid by comparison) and to slow down to enjoy some slow dishes traditionally cooked with it. (Like posole, a smoky Southwestern stew of dried roasted corn that, made right, can take all day.)

Paradoxically, sometimes the best way to rescue the most idiosyncratic local products and practices is to find a global market for them. This is what Slow Food means by "virtuous globalization," a simple but powerful idea that throws a wrench of complexity into the usual black-or-white arguments over free trade. It is no accident that food has emerged as a flash point in the free-trade debate; what we eat is a marker of our cultural identity, which is why threats to that identity, whether in the form of a new fast-food outlet or a genetically engineered crop, can excite such vehement reactions, as companies like McDonald's and Monsanto have discovered.

Certainly, the main tendency of globalism has been in the direction of the McDonald's ideal of "one world, one taste," but Slow Food makes a good case that globalism's power can also be exploited to save the local cultures most threatened by it. So a Piedmontese grower of a rare, wonderfully tasty but comparatively unproductive strain of wheat who can't find a local market can be, through Slow Food, hooked up with a company like Williams-Sonoma, which knows exactly where to find the affluent home bakers willing to pay a premium for a flour that makes such distinctive bread. One menu item at a time, Slow Food is demonstrating how global trade and mass communication can be turned into powerful tools for rescuing cultural and biological diversity—from precisely those perils of global trade and mass communication. Think of it as a form of economic jujitsu.

Carlo Petrini himself, a round, stubble-bearded Piedmontese in his 50s who looks like he knows how to enjoy himself at a table, has a genius for publicity that has been essential to the success of this strategy. He understands that the glamour that attaches to lavishly advertised global brands like McDonald's can be effectively countered only by creating a rival form of glamour. But how do you glamorize Iroquois corn flour or the rather scrawny Narragansett turkey? By recruiting great chefs to cook with these foods and extol their virtues. We live in an age when chefs wield unprecedented influence, and Slow Food has been quick to enlist them under its banner. Soon after Patrick Martins opened Slow Food's U.S. outpost in 2000, he invited Alice Waters, founder of Berkeley's Chez Panisse, to join the movement, and it wasn't long before much of America's culinary establishment has signed on too. Today, Slow Food USA has 10,000 members and 79 regional Convivia.

When merely promoting an endangered food isn't enough to save it from imminent extinction, Slow Food turns to its network of chefs and civilian members to organize a presidium. "Presidium" is Latin for "armed garrison," and this is as close to direct action as the movement gets. Take the case of America's endangered "heritage turkey" breeds. The Bourbon Red, the Narragansett, the Jersey Buff, and the Standard Bronze—the turkeys Americans ate for centuries—have all but succumbed to the aptly named and entirely flavorless Broad Breasted White. This is a turkey that has been so thoroughly industrialized (to produce lots of white meat fast) that it can no longer fly, survive outdoors, or reproduce without help. (Yep, the humongous breasts render conventional turkey sex impossible, so the birds must be artificially inseminated.) Today the U.S. turkey industry is a vast monoculture precariously perched on the beaks of these cosseted birds, which have driven older and more robust varieties to the edge of extinction—by one count only 3,800 breeding birds of these species survived at the millennium.

Early in 2002, Patrick Martins decided to organize a presidium to save the heritage turkey—Slow Food's term for the four old varieties it targeted. The New York office recruited a network of farmers to raise the birds from eggs it had persuaded hatcheries to produce. The organization guaranteed the farmers $3.50 a pound for its turkeys, advanced them start-up money for feed, and then set about finding restaurants and consumers willing to serve the birds for Thanksgiving. Some 5,000 orders came in, and in November 2002, Martins found himself at an Ohio slaughterhouse overseeing the processing of thousands of heritage turkeys, which wound up on the menus of restaurants all over America.

Not to mention on my own Thanksgiving table. I ordered a Narragansett from Pam Marshall, one of the farmers Slow Food had recruited, and paid it a couple of visits over the course of the summer. In 2002, Marshall grew Broad Breasted Whites and heritage turkeys side by side at her farm in Amenia, New York. She quickly learned why the BBWs ("mindless eating-and-shitting machines," she calls them) have prevailed in the marketplace: They were oven-ready by August, a full three months ahead of her Bourbon Reds and Narragansetts. The heritage birds took their sweet time getting up to slaughter weight, spending their days exploring Marshall's pastures, nibbling on clover and bugs, even doing a bit of flying now and then. At Thanksgiving, many of the turkeys were still small and flat of chest.

A handful of turkey buyers, including a few of the chefs, complained to Patrick Martins—they'd been promised 18-pounders, and only a few of the birds could hit that mark. But that's how it sometimes goes with slow food, Martins explained, shrugging his shoulders. "We don't call it Slow Food for nothing." These are, or were, living creatures, not factory-made products, and it is precisely our insistence on predictability and standardization, on quantity rather than quality, that has given us food that looks and tastes . . . well, as if it came from a factory.

As it happens, I had to dust off Martins' argument myself when I brought my heritage turkey to the table, a scene that would not have impressed Norman Rockwell. My Narragansett cut a fairly unprepossessing figure on the platter, being rather flat-chested and, um, bereft of both its legs. That's because Peter Hoffman, a Manhattan chef I'd consulted on cooking my turkey, had counseled me to braise the legs separately. "These birds actually use their legs," Hoffman explained. "The meat can be chewier than you're accustomed to." So I painstakingly removed the legs and braised them in chicken stock for several hours, all the while reminding my impatient self that this too was part of the whole slow-food experience.

The sniggering that initially greeted my slow turkey stopped the instant people had a chance to taste it. The leg and thigh meat in particular was delicious: rich, moist, and tender, with a flavor more reminiscent of duck than turkey. Indeed, simply by virtue of having a flavor, this represented a completely different order of turkey. Now I understood what turkey was like before the triumph of the Broad Breasted White, and why eating turkey had once been considered a great treat—heretofore one of the mysteries of life, as far as I could tell. It never occurred to my guests that by enjoying this Narragansett they were in some small way contributing to its survival, but of course they were: The Slow Food presidium in which I was taking part has already succeeded in nearly doubling the world population of heritage turkeys.

It all seemed too good to be true: that eating something this delicious could be a strategy for preserving biodiversity and that the pleasure we took in doing so could itself constitute a small but meaningful political act. For pleasure itself is all but extinct in American environmentalism (not to mention American eating), and pleasure is part of what Slow Food aims to redeem, by demonstrating that, at least when it comes to the politics of food, the best choice is often the tastiest. Eco-gastronomy isn't going to save the world, but if it can bring politics and pleasure together on the American plate, the Vesuvian apricot and Delaware Bay oyster won't be the only species to benefit.

WILLIAM LANGEWIESCHE

Columbia's *Last Flight*

FROM *THE ATLANTIC MONTHLY*

S pace flight is known to be a risky business, but during the minutes before
dawn February 1, as the doomed shuttle *Columbia* began to descend into
the upper atmosphere over the Pacific Ocean, only a handful of people—
a few engineers deep inside of NASA—worried that the vehicle and its seven
souls might actually come to grief. It was the responsibility of NASA's managers to hear those suspicions, and from top to bottom they failed. After the
fact, that's easy to see. But in fairness to those whose reputations have now been
sacrificed, seventeen years and eighty-nine shuttle flights had passed since the
Challenger explosion, and within the agency a new generation had risen that
was smart, perhaps, but also unwise—confined by NASA's walls and routines,
and vulnerable to the self-satisfaction that inevitably had set in.

Moreover, this mission was a yawn—a low-priority "science" flight forced
onto NASA by Congress and postponed for two years because of a more pressing schedule of construction deliveries to the International Space Station. The
truth is, it had finally been launched as much to clear the books as to add to
human knowledge, and it had gone nowhere except into low Earth orbit,
around the globe every ninety minutes for sixteen days, carrying the first Israeli
astronaut, and performing a string of experiments, many of which, like the
shuttle program itself, seemed to suffer from something of a make-work character—the examination of dust in the Middle East (by the Israeli, of course);

the ever popular ozone study; experiments designed by schoolchildren in six countries to observe the effect of weightlessness on spiders, silkworms, and other creatures; an exercise in "astroculture" involving the extraction of essential oils from rose and rice flowers, which was said to hold promise for new perfumes; and so forth. No doubt some good science was done too—particularly pertaining to space flight itself—though none of it was so urgent that it could not have been performed later, under better circumstances, in the underbooked International Space Station. The astronauts aboard the shuttle were smart and accomplished people, and they were deeply committed to human space flight and exploration. They were also team players, by intense selection, and nothing if not wise to the game. From orbit one of them had radioed, "The science we're doing here is great, and it's fantastic. It's leading-edge." Others had dutifully reported that the planet seems beautiful, fragile, and borderless when seen from such altitudes, and they had expressed their hopes in English and Hebrew for world peace. It was Miracle Whip on Wonder Bread, standard NASA fare. On the ground so little attention was being paid that even the radars that could have been directed upward to track the *Columbia*'s re-entry into the atmosphere—from Vandenberg Air Force Base, or White Sands Missile Range—were sleeping. As a result, no radar record of the breakup exists—only of the metal rain that drifted down over East Texas, and eventually came into the view of air-traffic control.

Along the route, however, stood small numbers of shuttle enthusiasts, who had gotten up early with their video cameras and had arrayed themselves on hills or away from city lights to record the spectacle of what promised to be a beautiful display. The shuttle came into view, on track and on schedule, just after 5:53 Pacific time, crossing the California coast at about 15,000 mph in the superthin air 230,000 feet above the Russian River, northwest of San Francisco. It was first picked up on video by a Lockheed engineer in suburban Fairfield, who recorded a bright meteor passing almost directly overhead, not the shuttle itself but the sheath of hot gases around it, and the long, luminous tail of ionized air known as plasma. Only later, after the engineer heard about the accident on television, did he check his tape and realize that he had recorded what appeared to be two pieces coming off the Columbia in quick succession, like little flares in its wake. Those pieces were recorded by others as well, along with the third, fourth, and fifth "debris events" that are known to have occurred during the sixty seconds that it took the shuttle to cross California. From the top of Mount Hamilton, southeast of San Francisco, another engineer, the former president of the Peninsula Astronomical Society, caught all five events on tape but, again, did not realize it until afterward. He later said, "I'd seen four re-entries before this one. When we saw it, we did note that it was a little brighter and a little bit whiter in color than it normally is. It's normally a pink-magenta color. But you know, it wasn't so different that it really flagged us as something wrong. With the naked eye we didn't see the particles coming off."

One minute after the *Columbia* left California, as it neared southwestern Utah, the trouble was becoming more obvious to observers on the ground. There had been a bright flash earlier over Nevada, and now debris came off that was large enough to cause multiple secondary plasma trails. North of the Grand Canyon, in Saint George, Utah, a man and his grown son climbed onto a ridge above the country hospital, hoping for the sort of view they had seen several years before, of a fireball going by. It was a sight they remembered as "really neat." This time was different, though. The son, who was videotaping, started yelling, "Jesus, Dad, there's stuff falling off!" and the father saw it too, with his naked eyes.

The *Columbia* was flying on autopilot, as is usual, and though it continued to lay flares in its wake, the astronauts aboard remained blissfully unaware of the trouble they were in. They passed smoothly into dawn above the Arizona border, and sailed across the Navajo reservation and on over Albuquerque, before coming to the Texas Panhandle on a perfect descent profile, slowing through 13,400 mph at 210,000 feet five minutes after having crossed the California coastline. Nineteen seconds later, at 7:58:38 central time, they got the first sign of something being a little out of the ordinary: it was a cockpit indication of low tire pressures on the left main landing gear. This was not quite a trivial matter. A blown or deflated main tire would pose serious risks during the roll-out after landing, including loss of lateral control and the possibility that the nose would slam down, conceivably leading to a catastrophic breakup on the ground. These scenarios were known, and had been simulated and debated in the inner world of NASA, leading some to believe that the best of the imperfect choices in such a case might be for the crew to bail out—an alternative available only below 30,000 feet and 220 mph of dynamic airspeed.

Nonetheless, for *Columbia*'s pilots it was reasonable to assume for the moment that the indication of low pressure was due to a problem with the sensors rather than with the tires themselves, and that the teams of Mission Control engineers at NASA's Johnson Space Center, in Houston, would be able to sort through the mass of automatically transmitted data—the so-called telemetry, which was far more complete than what was available in the cockpit—and to draw the correct conclusion. The reverse side of failures in a machine as complex as the shuttle is that most of them can be worked around, or turn out to be small. In other words, there was no reason for alarm. After a short delay the *Columbia*'s commander, Rick Husband, calmly radioed to Mission Control, "And, ah, Houston . . ." Sheathed in hot atmospheric gases, the shuttle was slowing through 13,100 mph at 205,000 feet.

Houston did not clearly hear the call.

With the scheduled touchdown now only about fifteen minutes ahead, it was a busy time at Mission Control. Weather reports were coming in from the landing site at the Kennedy Space Center, in Florida. Radar tracking of the shuttle, like the final accurate ground-based navigation, had not yet begun. Sitting at their specialized positions, and monitoring the numbers displayed on the consoles, a few of the flight controllers had begun to sense, just barely, that something was going seriously wrong. The worry was not quite coherent yet. One of the controllers later told me that it amounted to an inexplicable bad feeling in his gut. But it was undeniable nonetheless. For the previous few minutes, since about the time when the shuttle had passed from California to Nevada, Jeff Kling, an engineer who was working the mechanical-systems position known as MMACS (pronounced Macs), had witnessed a swarm of erratic indications and sensor failures. The pattern was disconcerting because of the lack of common circuitry that could easily explain the pattern of such failures—a single box that could be blamed.

Kling had been bantering good-naturedly on an intercom with one of his team, a technician sitting in one of the adjoining back rooms and monitoring the telemetry, when the technician noted a strange failure of temperature transducers on a hydraulic return line. The technician said, "We've had some hydraulic 'ducers go off-scale low."

Kling had seen the same indications. He said, "Well, I guess!"

The technician said, "What in the world?"

Kling said, "This is not funny. On the left side."

The technician confirmed, "On the left side . . ."

Now Kling got onto the main control-room intercom to the lead controller on duty, known as the flight director, a man named Leroy Cain. In the jargon-laced language of the control room Kling said, "Flight, Macs."

Cain said, "Go ahead, Macs."

"FYI, I've just lost four separate temperature transducers on the left side of the vehicle, hydraulic return temperatures. Two of them on system one, and one in each of systems two and three."

Cain said, "Four hyd return temps?"

Kling answered, "To the left outboard and left inboard elevon."

"Okay, is there anything common to them? DSC or MDM or anything? I mean, you're telling me you lost them all at exactly the same time?"

"No, not exactly. They were within probably four or five seconds of each other."

Cain struggled to assess the meaning. "Okay, where are those . . . where is that instrumentation located?"

Kling continued to hear from his back-room team. He said, "All four of them are located in the aft part of the left wing, right in front of the elevons . . . elevon actuators. And there is no commonality."

Cain repeated, "No commonality."

But all the failing instruments were in the left wing. The possible significance of this was not lost on Cain: during the launch a piece of solid foam had broken off from the shuttle's external fuel tank, and at high speed had smashed into the left wing; after minimal consideration the shuttle program managers (who stood above Mission Control in the NASA hierarchy) had dismissed the incident as essentially unthreatening. Like almost everyone else at NASA, Cain had taken the managers at their word—and he still did. Nonetheless, the strange cluster of left-wing failures was an ominous development. Kling had more-specific reasons for concern. In a wonkish, engineering way he had discussed with his team the telemetry they might observe if a hole allowed hot gases into the wing during re-entry, and had come up with a profile eerily close to what was happening now. Still, he maintained the expected detachment.

Cain continued to worry the problem. He asked for reassurance from his "guidance, navigation, and control" man, Mike Sarafin. "Everything look good to you, control and rates and everything is normal, right?"

Sarafin said, "Control's been stable through the rolls that we've done so far, Flight. We have good trims. I don't see anything out of the ordinary."

Cain directed his attention back to Kling: "All other indications for your hydraulic systems indications are good?"

"They're all good. We've had good quantities all the way across."

Cain said, "And the other temps are normal?"

"The other temps are normal, yes, sir." He meant only those that the telemetry allowed him to see.

Cain said, "And when you say you lost these, are you saying they went to zero . . ."

"All four of them are off-scale low."

". . . or off-scale low?"

Kling said, "And they were all staggered. They were, like I said, within several seconds of each other."

Cain said, "Okay."

But it wasn't okay. Within seconds the *Columbia* had crossed into Texas and the left-tire-pressure indications were dropping, as observed also by the cockpit crew. Kling's informal model of catastrophe had predicted just such indications, whether from blown tires or wire breaks. The end was now coming very fast.

Kling said, "Flight, Macs."

Cain said, "Go."

"We just lost tire pressure on the left outboard and left inboard, both tires."

Cain said, "Copy."

At that moment, twenty-three seconds after 7:59 local time, the Mission Control consoles stopped receiving telemetry updates, for reasons unknown. The astronaut sitting beside Cain, and serving as the Mission Control communicator, radioed, "And *Columbia*, Houston, we see your tire-pressure messages, and we did not copy your last call."

At the same time, on the control-room intercom, Cain was talking again to Kling. He said, "Is it instrumentation, Macs? Gotta be."

Kling said, "Flight, Macs, those are also off-scale low."

From the speeding shuttle Rick Husband—Air Force test pilot, religious, good family man, always wanted to be an astronaut—began to answer the communicator. He said, "Roger, ah," and was cut off on a word that began with "buh . . ."

It turned out to be the *Columbia*'s last voice transmission. Brief communication breaks, however, are not abnormal during re-entries, and this one raised no immediate concern in Houston.

People on the ground in Dallas suddenly knew more than the flight controllers in Houston. Four seconds after eight they saw a large piece leave the orbiter and fall away. The shuttle was starting to come apart. It continued intermittently to send telemetry, which though not immediately displayed at Mission Control was captured by NASA computers and later discovered; the story it told was that multiple systems were failing. In quick succession two additional chunks fell off.

Down in the control room Cain said, "And there's no commonality between all these tire-pressure instrumentations and the hydraulic return instrumentations?"

High in the sky near Dallas the *Columbia*'s main body began to break up. It crackled and boomed, and made a loud rumble.

Kling said, "No, sir, there's not. We've also lost the nose-gear down talkback, and right-main-gear down talkback."

"Nose-gear and right-main-gear down talkbacks?"

"Yes, sir."

At Fort Hood, Texas, two Dutch military pilots who were training in an Apache attack helicopter locked on to the breakup with their optics and video-taped three bright objects—the main rocket engines—flying eastward in formation, among other, smaller pieces and their contrails.

Referring to the loss of communications, one minute after the main-body breakup, Laura Hoppe, the flight controller responsible for the communications systems, said to Cain, "I didn't expect, uh, this bad of a hit on comm."

Cain asked another controller about a planned switchover to a ground-based radio ahead, "How far are we from UHF? Is that two-minute clock good?"

Kling, also, was hanging on to hope. He said, "Flight, Macs."

Cain said, "Macs?"

Kling said, "On the tire pressures, we did see them go erratic for a little bit before they went away, so I do believe it's instrumentation."

"Okay."

At about that time the debris began to hit the ground. It fell in thousands of pieces along a swath ten miles wide and 300 miles long, across East Texas and into Louisiana. There were many stories later. Some of the debris whistled down through the leaves of trees and smacked into a pond where a man was fishing. Another piece went right through a backyard trampoline, evoking a mother's lament: "Those damned kids . . ." Still another piece hit the window of a moving car, startling the driver. The heaviest parts flew the farthest. An 800-pound piece of engine hit the ground in Fort Polk, Louisiana, doing 1,400 mph. A 600-pound piece landed nearby. Thousands of people began to call in, swamping the 911 dispatchers with reports of sonic booms and metal falling out of the sky. No one, however, was hit. This would be surprising were it not for the fact, so visible from above, that the world is still a sparsely populated place.

In Houston the controllers maintained discipline, and continued preparing for the landing, even as they received word that the Merritt Island radar, in Florida, which should by now have started tracking the inbound craft, was picking up only false targets. Shuttles arrive on time or they don't arrive at all. But, repeatedly, the communicator radioed, "*Columbia*, Houston, UHF comm check," as if he might still hear a reply. Then, at thirteen minutes past the hour, precisely when the *Columbia* should have been passing overhead the runway before circling down for a landing at the Kennedy Space Center, a phone call came in from an off-duty controller who had just seen a video broadcast by a

Dallas television station of multiple contrails in the sky. When Cain heard the news, he paused, and then put the contingency plan into effect. To the ground-control officer he said, "GC, Flight."

"Flight, GC."

"Lock the doors."

"Copy."

The controllers were stunned, but lacked the time to contemplate the horror of what had just happened. Under Cain's direction they set about collecting numbers, writing notes, and closing out their logs, for the investigation that was certain to follow. The mood in the room was somber and focused. Only the most basic facts were known: the *Columbia* had broken up at 200,000 feet doing 12,738 mph, and the crew could not possibly have survived. Ron Ditte-more, the shuttle program manager, would be talking to reporters later that day, and he needed numbers and information. At some point sandwiches were brought in and consumed. Like the priests who harvest faith at the bedsides of the dying, grief counselors showed up too, but they were not much used.

Cain insisted on control-room discipline. He said, "No phone calls off site outside of this room. Our discussions are on these loops—the recorded DVIS loops only. No data, no phone calls, no transmissions anywhere, into or out."

Later this was taken by some critics to be a typical NASA reaction—insular, furtive, overcontrolling. And it may indeed have reflected certain aspects of what had become of the agency's culture. But it was also, more simply, a rule-book procedure meant to stabilize and preserve the crucial last data. The room was being frozen as a crime scene might be. Somewhere inside NASA something had obviously gone very wrong—and it made sense to start looking for the evidence here and now.

LESS THAN AN HOUR LATER, at 10:00 AM eastern time, a retired four-star admiral named Hal Gehman met his brother at a lawyer's office in Williamsburg, Virginia. At the age of sixty, Gehman was a tall, slim, silver-haired man with an unlined face and soft eyes. Dressed in civilian clothes, standing straight but not stiffly so, he had an accessible, unassuming manner that contrasted with the rank and power he had achieved. After an inauspicious start as a mediocre engineering student in the Penn State Naval ROTC program ("Top four fifths of the class," he liked to say), he had skippered a patrol boat through the thick of the Vietnam War and gone on to become an experienced sea captain, the commander of a carrier battle group, vice-chief of the Navy, and finally NATO Atlantic commander and head of the U.S. Joint

Forces Command. Upon his retirement, in 2000, from the sixth-ranked position in the U.S. military, he had given all that up with apparent ease. He had enjoyed a good career in the Navy, but he enjoyed his civilian life now too. He was a rare sort of man—startlingly intelligent beneath his guileless exterior, personally satisfied, and quite genuinely untroubled. He lived in Norfolk in a pleasant house that he had recently remodeled; he loved his wife, his grown children, his mother and father, and all his siblings. He had an old Volkswagen bug convertible, robin's-egg blue, that he had bought from another admiral. He had a modest thirty-four-foot sloop, which he enjoyed sailing in the Chesapeake, though its sails were worn out and he wanted to replace its icebox with a twelve-volt refrigeration unit. He was a patriot, of course, but not a reactionary. He called himself a fiscal conservative and a social moderate. His life as he described it was the product of convention. It was also the product of a strict personal code. He chose not to work with any company doing business with the Department of Defense. He liked power, but understood its limitations. He did not care to be famous or rich. He represented the American establishment at its best.

In the lawyer's office in Williamsburg his brother told him that the *Columbia* had been lost. Gehman had driven there with his radio off and so he had not heard. He asked a few questions, and absorbed the information without much reaction. He did not follow the space program and, like most Americans, had not been aware that a mission was under way. He spent an hour with the lawyer on routine family business. When he emerged, he saw that messages had been left on his cell phone, but because the coverage was poor, he could not retrieve them; only later, while driving home on the interstate, was he finally able to connect. To his surprise, among mundane messages he found an urgent request to call the deputy administrator of NASA, a man he had not heard of before, named Fred Gregory. Like a good American, Gehman made the call while speeding down the highway. Gregory, a former shuttle commander, said, "Have you heard the news?"

Gehman said, "Only secondhand."

Gregory filled him in on what little was known, and explained that part of NASA's contingency plan, instituted after the *Challenger* disaster of 1986, was the activation of a standing "interagency" investigation board. By original design the board consisted of seven high-ranking civilian and military officials who were pre-selected mechanically on the basis of job titles—the institutional slots that they filled. For the *Columbia*, the names were now known: the board would consist of three Air Force generals, John Barry, Kenneth Hess, and Duane Deal; a Navy admiral, Stephen Turcotte; a NASA research director,

G. Scott Hubbard; and two senior civil-aviation officials, James Hallock and Steven Wallace. Though only two of these men knew much about NASA or the space shuttle, in various ways each of them was familiar with the complexities of large-scale, high-risk activities. Most of them also had strong personalities. To be effective they would require even stronger management. Gregory said that it was NASA's administrator, Sean O'Keefe, who wanted Gehman to come in as chairman to lead the work. Gehman was not immune to the compliment, but he was cautious. He had met O'Keefe briefly years before, but did not know him. He wanted to make sure he wasn't being suckered into a NASA sideshow.

O'Keefe was an able member of Washington's revolving-door caste, a former congressional staffer and budget specialist—and a longtime protégé of Vice President Dick Cheney—who through the force of his competence and Republican connections had briefly landed the position of Secretary of the Navy in the early 1990s. He had suffered academic banishment through the Clinton era, but under the current administration had re-emerged as a deputy at the Office of Management and Budget, where he had been assigned to tackle the difficult problem of NASA's cost overruns and lack of delivery, particularly in the Space Station program. It is hard to know what he thought when he was handed the treacherous position of NASA administrator. Inside Washington, NASA's reputation had sunk so low that some of O'Keefe's former congressional colleagues snickered that Cheney was trying to kill his own man off. But O'Keefe was not a space crusader, as some earlier NASA administrators had been, and he was not about to pick up the fallen banners of the visionaries and try to lead the way forward; he was a tough, level-headed money man, grounded in the realities of Washington, DC, and sent in on a mission to bring discipline to NASA's budget and performance before moving on. NASA's true believers called him a carpetbagger and resented the schedule pressures that he brought to bear, but in fairness he was a professional manager, and NASA needed one.

O'Keefe had been at NASA for just over a year when the *Columbia* self-destructed. He was in Florida standing at the landing site beside one of his deputies, a former shuttle commander named William Readdy. At 9:05 eastern time, ten minutes before the scheduled landing, Readdy got word that communications with the shuttle, which had been lost, had not been re-established; O'Keefe noticed that Readdy's face went blank. At 9:10 Readdy opened a book to check a time sequence. He said, "We should have heard the sonic booms by now. There's something really wrong." By 9:29 O'Keefe had activated the full-blown contingency plan. When word got to the White House, the executive staff ducked quickly into defensive positions: President Bush would grieve alongside the families and say the right things about carrying on, but rather than involving himself by appointing an independent presidential commission, as Ronald Reagan had in response to the *Challenger* accident, he would keep his distance by expressing faith in NASA's ability to find the cause. In other words, this baby was going to be dropped squarely onto O'Keefe's lap. The White House approved Gehman's appointment to lead what would essentially be NASA's investigation—but O'Keefe could expect little further communication. There was a chance that the President would not even want to receive the final report directly but would ask that it be deposited more discreetly in the White House in-box. He had problems bigger than space on his mind.

Nonetheless, that morning in his car Gehman realized that even with a lukewarm White House endorsement, the position that NASA was offering, if handled correctly, would allow for a significant inquiry into the accident. Gregory made it clear that Gehman would have the full support of NASA's engineers and technical resources in unraveling the physical mysteries of the accident—what actually had happened to the *Columbia* out there in its sheath of fire at 200,000 feet. Moreover, Gehman was confident that if the investigation had to go further, into why this accident had occurred, he had the experience necessary to sort through the human complexities of NASA and emerge with useful answers that might result in reform. This may have been overconfident of him, and to some extent Utopian, but it was not entirely blind: he had been through big investigations before, most recently two years earlier, just after leaving the Navy, when he and a retired Army general named William Crouch had led an inquiry into the loss of seventeen sailors aboard the USS *Cole,* the destroyer that was attacked and nearly sunk by suicide terrorists in Yemen in October of 2000. Their report found fundamental errors in the functioning of the military command structure, and issued recommendations (largely classified) that are in effect today. The success of the *Cole* investigation was one of the arguments that Gregory used on him now. Gehman did not disagree, but he wanted to be very clear. He said, "I know you've got a piece of paper in front of you. Does it say that I'm not an aviator?"

Gregory said, "We don't need an aviator here. We need an investigator."

And so, driving down the highway to Norfolk, Gehman accepted the job. When he got home, he told his wife that he was a federal employee again and that there wouldn't be much sailing in the spring. That afternoon and evening, as the faxes and phone calls came in, he began to exercise control of the process, if only in his own mind, concluding that the board's charter as originally written by NASA would have to be strengthened and expanded, and that its name should immediately be changed from the absurd International Space

Station and Space Shuttle Mishap Interagency Investigations Board (the ISSSSMIIB) to the more workable *Columbia* Accident Investigation Board, or CAIB, which could be pronounced in one syllable, as Cabe.

NASA initially did not resist any of his suggestions. Gregory advised Gehman to head to Barksdale Air Force Base, in Shreveport, Louisiana, where the wreckage was being collected. As Gehman began to explore airline connections, word came that a NASA executive jet, a Gulfstream, would be dispatched to carry him, along with several other board members, directly to Barksdale. The jet arrived in Norfolk on Sunday afternoon, the day after the accident. One of the members already aboard was Steven Wallace, the head of accident investigations for the FAA. Wallace is a second-generation pilot, an athletic, tightly wound man with wide experience in government and a skeptical view of the powerful. He later told me that when Gehman got on the airplane, he was dressed in a business suit, and that, having introduced himself, he explained that they might run into the press, and if they did, he would handle things. This raised some questions about Gehman's motivations (and indeed Gehman turned out to enjoy the limelight), but as Wallace soon discovered, grandstanding was not what Gehman was about. As the Gulfstream proceeded toward Louisiana, Gehman rolled up his sleeves and, sitting at the table in the back of the airplane, began to ask for the thoughts and perspectives of the board members there—not about what might have happened to the *Columbia* but about how best to find out. It was the start of what would become an intense seven-month relationship. It was obvious that Gehman was truly listening to the ideas, and that he was capable of integrating them quickly and productively into his own thoughts. By the end of the flight even Wallace was growing impressed.

But Gehman was in some ways also naive, formed as he had been by investigative experience within the military, in which much of the work proceeds behind closed doors, and conflict of interest is not a big concern. The *Columbia* investigation, he discovered, was going to be a very different thing. Attacks against the CAIB began on the second day, and by midweek, as the board moved from Shreveport to Houston to set up shop, they showed no signs of easing. Congress in particular was thundering that Gehman was a captive investigator, that his report would be a whitewash, and that the White House should replace the CAIB with a *Challenger*-style presidential commission. This came as a surprise to Gehman, who had assumed that he could just go about his business but who now realized that he would have to accommodate these concerns if the final report was to have any credibility at all. Later he said to me, "I didn't go in thinking about it, but as I began to hear the independence thing—'You can't have a panel appointed by NASA investigating itself!'—I realized I'd better deal with Congress." He did this at first mainly by listening on the phone. "They told me what I had to do to build my credibility. I didn't invent it—they told me. They also said, 'We hate NASA. We don't trust them. Their culture is no good. And their cost accounting is no good.' And I said, 'Okay.' "

More than that, Gehman came to realize that it was the elected representatives in Congress—and neither O'Keefe nor NASA—who constituted the CAIB's real constituency, and that their concerns were legitimate. As a result of this, along with a growing understanding of the depth and complexity of the work at hand, he forced through a series of changes, establishing a congressional-liaison office, gaining an independent budget (ultimately of about $20 million), wresting the report from O'Keefe's control, rewriting the stated mission to include the finding of "root causes and circumstances," and hiring an additional five board members, all civilians of unimpeachable reputation: the retired Electric Boat boss Roger Tetrault, the former astronaut Sally Ride, the Nobel-laureate physicist Douglas Osheroff, the aerodynamicist and former Air Force Secretary Sheila Widnall, and the historian and space-policy expert John Logsdon. Afterward, the loudest criticism faded away. Still, Gehman's political judgment was not perfect. He allowed the new civilian members to be brought on through the NASA payroll (at prorated annual salaries of $134,000)—a strange lapse under the circumstances, and one that led to superficial accusations that the CAIB remained captive. The *Orlando Sentinel* ran a story about the lack of public access to the CAIB's interviews under the ambiguous headline "BOARD PAID TO ENSURE SECRECY." The idea evoked laughter among some of the investigators, who knew the inquiry's direction. But unnecessary damage was done.

Equally unnecessary was Gehman's habit of referring to O'Keefe as "Sean," a clubbish mannerism that led people to conclude, erroneously, that the two men were friends. In fact their relationship was strained, if polite. Gehman told me that he had never asked for the full story behind his selection on the morning of the accident—maybe because it would have been impossible to know the unvarnished truth. Certainly, though, O'Keefe had had little opportunity to contemplate his choice. By quick view Gehman was a steady hand and a good establishment man who could lend the gravitas of his four stars to this occasion; he was also, of course, one of the men behind the *Cole* investigation. O'Keefe later told me that he had read the *Cole* report during his stint as a professor, but that he remembered it best as the subject of a case study presented by one of his academic colleagues as an example of a narrowly focused investigation that, correctly, had not widened beyond its original mandate. This was true, but a poor predictor of Gehman as a man. His *Cole* investigation had not widened (for instance, into assigning individual blame) for the simple reason that other investigations, by the Navy and the FBI, were already covering that ground. Instead, Gehman and Crouch had gone deep, and relentlessly so. The result was a document that bluntly questioned current American dogma, identified arrogance in the command structure, and critiqued U.S. military assumptions about the terrorist threat. The tone was frank. For example, while expressing understanding of the diplomatic utility of labeling terrorists as "criminals," the report warned against buying into that language, or into the parallel idea that these terrorists were "cowards." When, later, I expressed my surprise at his freedom of expression, Gehman did not deny that people have recently been decried as traitors for less. But freedom of expression was clearly his habit: he spoke to me just as openly about the failures of his cherished Navy, of Congress, and increasingly of NASA.

When I mentioned this character trait to one of the new board members, Sheila Widnall, she laughed and said she'd seen it before inside the Pentagon, and that people just didn't understand the highest level of the U.S. military. These officers are indeed the establishment, she said, but they are so convinced of the greatness of the American construct that they will willingly tear at its components in the belief that its failures can be squarely addressed. Almost all of the current generation of senior leaders have also been through the soul-searching that followed the defeat in Vietnam.

O'Keefe had his own understanding of the establishment, and it was probably sophisticated, but he clearly did not anticipate Gehman's rebellion. By the end of the second week, as Gehman established an independent relationship with Congress and began to break through the boundaries initially drawn by NASA, it became clear that O'Keefe was losing control. He maintained a brave front of wanting a thorough inquiry, but it was said that privately he was angry. The tensions came to the surface toward the end of February, at about the same time that Gehman insisted, over O'Keefe's resistance, that the full report ultimately be made available to the public. The CAIB was expanding to a staff of about 120 people, many of them professional accident investigators and technical experts who could support the core board members. They were working seven days a week out of temporary office space in the sprawling wasteland of South Houston, just off the property of the Johnson Space Center. One morning several of the board members came in to see Gehman, and warned him that the CAIB was headed for a "shipwreck."

Gehman knew what they meant. In the days following the accident O'Keefe had established an internal Mishap Investigation Team, whose job was to work closely with the CAIB, essentially as staff, and whose members—bizarrely—included some of the decision-makers most closely involved with the Columbia's final flight. The team was led by Linda Ham, a razor-sharp manager in the shuttle program, whose actions during the flight would eventually be singled out as an egregious example of NASA's failings. Gehman did not know that yet, but it dawned on him that Ham was in a position to filter the inbound NASA reports, and he remembered a recent three-hour briefing that she had run with an iron hand, allowing little room for spontaneous exploration. He realized that she and the others would have to leave the CAIB, and he wrote a careful letter to O'Keefe in Washington, requesting their immediate removal. It is a measure of the insularity at the Johnson Space Center that NASA did not gracefully acquiesce. Ham and another manager, Ralph Roe, in particular reacted badly. In Gehman's office, alternately in anger and tears, they refused to leave, accusing Gehman of impugning their integrity and asking him how they were supposed to explain their dismissal to others. Gehman suggested to them what Congress had insisted to him—that people simply cannot investigate themselves. Civics 101. Once stated, it seems like an obvious principle.

O'Keefe had a master's degree in public administration, but he disagreed. It was odd. He had not been with the agency long enough to be infected by its insularity, and as he later promised Congress, he was willing—no, eager—to identify and punish any of his NASA subordinates who could be held responsible for the accident. Nonetheless, he decided to defy Gehman, and he announced that his people would remain in place. It was an ill-considered move. Gehman simply went public with his letter, posting it on the CAIB Web site. Gehman understood that O'Keefe felt betrayed—"stabbed in the back" was the word going around—but NASA had left him no choice. O'Keefe surrendered. Ham and the others were reassigned, and the Mishap Investigation Team was disbanded, replaced by NASA staffers who had not been involved in the Columbia's flight and would be more likely to cooperate with the CAIB's investigators. The board was never able to overcome completely the whiff of collusion that had accompanied its birth, but Gehman had won a significant fight, even if it meant that he and "Sean" would not be friends.

THE SPACE SHUTTLE is the most audacious flying machine ever built, an engineering fantasy made real. Before each flight it stands vertically on the launch pad at the Kennedy Space Center, as the core component of a rocket assembly 184 feet tall. The shuttle itself, which is also known as the orbiter, is a winged vehicle roughly the size of a DC-9, with three main rocket engines in the tail, a large unpressurized cargo bay in the midsection, and a cramped two-level crew compartment in the nose. It is attached to a huge external tank containing liquid fuel for the three main engines. That tank in turn is attached to two solid-fuel rockets, known as boosters, which flank the assembly and bear its full weight on the launch pad. Just before the launch, the weight is about 4.5 million pounds, 90 percent of which is fuel. It is a dramatic time, ripe with anticipation; the shuttle vents vapors like a breathing thing; the ground crews pull away until finally no one is left; the air seems unusually quiet.

Typically there are seven astronauts aboard. Four of them sit in the cockpit, and three on the lower level, in the living quarters known as the mid-deck. Because of the shuttle's vertical position, their seats are effectively rotated backward 90 degrees, so they are sitting on their backs, feeling their own weight in a way that tends to emphasize gravity's pull. At the front of the cockpit, positioned closer to the instrument panel than is necessary for the typical astronaut's six-foot frame, the commander and the pilot can look straight ahead into space. They are highly trained. They know exactly what they are getting into. Sometimes they have waited years for this moment to arrive.

The launch window may be just a few minutes wide. It is ruled by orbital mechanics, and defined by the track and position of the destination—usually now the unfinished International Space Station. Six seconds before liftoff the three main engines are ignited and throttled up to 100 percent power, producing more than a million pounds of thrust. The shuttle responds with what is known as "the twang," swaying several feet in the direction of the external tank and then swaying back. This is felt in the cockpit. The noise inside is not very loud. If the computers show that the main engines are operating correctly, the solid rocket boosters ignite. The boosters are ferocious devices—the same sort of monsters that upon failure blew the *Challenger* apart. Each of them produces three million pounds of thrust. Once ignited, they cannot be shut off or throttled back. The shuttle lifts off. It accelerates fast enough to clear the launch tower doing about 100 mph, though it is so large that seen from the outside, it appears to be climbing slowly.

The flying is done entirely by autopilot unless something goes wrong. Within seconds the assembly rotates and aims on course, tilting slightly off the vertical and rolling so that the orbiter is inverted beneath the external tank. Although the vibrations are heavy enough to blur the instruments, the acceleration amounts to only about 2.5 Gs—a mild sensation of heaviness pressing the astronauts back into their seats. After about forty seconds the shuttle accelerates through Mach 1, 760 mph, at about 17,000 feet, climbing nearly straight up. Eighty seconds later, with the shuttle doing about 3,400 mph and approaching 150,000 feet, the crew can feel the thrust from the solid rocket boosters begin to tail off. Just afterward, with a bright flash and a loud explosion heard inside the orbiter, the rocket boosters separate from the main tank; they continue to travel upward on a ballistic path to 220,000 feet before falling back and parachuting into the sea. Now powered by the main engines alone, the ride turns smooth, and the forces settle down to about 1 G.

One pilot described the sensations to me on the simplest level. He said, "First it's like, 'Hey, this is a rough ride!' and then, 'Hey, I'm on an electric train!' and then, 'Hey, this train's starting to go pretty darned fast!' " Speed is the ultimate goal of the launch sequence. Having climbed steeply into ultra-thin air, the shuttle gently pitches over until it is flying nearly parallel to Earth, inverted under the external tank, and thrusting at full power. Six minutes after launch, at about 356,000 feet, the shuttle is doing around 9,200 mph, which is fast, but only about half the speed required to sustain an orbit. It therefore begins a shallow dive, during which it gains speed at the rate of 1,000 mph every twenty seconds—an acceleration so fast that it presses the shuttle against its 3 G limit, and the engines have to be briefly throttled back. At 10,300 mph the shuttle rolls to a head-up position. Passing through 15,000 mph, it begins to climb again, still accelerating at 3 Gs, until, seconds later, in the near vacuum of space, it achieves orbital velocity, or 17,500 mph. The plumes from the main engines wrap forward and dance across the cockpit windows, making light at night like that of Saint Elmo's fire. Only eight and a half minutes have passed since the launch. The main engines are extinguished, and the external tank is jettisoned. The shuttle is in orbit. After further maneuvering it assumes its standard attitude, flying inverted in relation to Earth and tail first as it proceeds around the globe.

For the astronauts aboard, the uphill flight would amount to little more than an interesting ride were it not for the possibility of failures. That possibility, however, is very real, and as a result the launch is a critical and complicated operation, demanding close teamwork, tight coordination with Mission Control, and above all extreme concentration—a quality often confused with coolness under fire. I was given a taste of this by an active shuttle commander named Michael Bloomfield, who had me strap in beside him in NASA's full-motion simulator in Houston, and take a realistic run from the launch pad into space. Bloomfield is a former Air Force test pilot who has flown three shuttle missions. He had been assigned to assist the CAIB, and had been watching the investigation with mixed emotions—hopeful that some effects might be positive, but concerned as well that the inquiry might veer into formalism without sufficiently taking into account the radical nature of space flight, or the basic truth that every layer of procedure and equipment comes at a cost, often unpredictable. Bloomfield called this the "risk versus risk" tradeoff, and made it real not by defending NASA against specific criticisms but by immersing me, a pilot myself, in the challenges of normal operations.

Much of what he showed me was of the what-if variety, the essence not only of simulator work but also of the crew's real-world thinking. For instance, during the launch, as the shuttle rockets upward on autopilot, the pilots and flight controllers pass through a succession of mental gates, related to various combinations of main-engine failures, at various altitudes and speeds. The options and resulting maneuvers are complicated, ranging from a quick return to the launch site, to a series of tight arrivals at select runways up the eastern seaboard, to transatlantic glides, and finally even an "abort into orbit"—an escape route used by a *Challenger* crew in 1985 after a single main-engine failure. Such failures allow little time to make the right decision. As Bloomfield and I climbed away from Earth, tilted onto our backs, he occasionally asked the operators to freeze the simulation so that he could unfold his thoughts to me. Though the choices were clear, the relative risks were rarely so obvious. It was a deep view into the most intense sort of flying.

After we arrived in space, we continued to talk. One of the gates for engine failure during the climb to the Space Station stands at Mach 21.8 (14,900 mph), the last point allowed for a "high energy" arrival into Gander, Newfoundland, and the start of the emergency transatlantic track for Shannon, Ireland. An abort at that point provides no easy solution. The problem with Gander is how to bleed off excess energy before the landing (Bloomfield called this "a take-all-your-brain-cells type of flying"), whereas the problem with Shannon is just the opposite—how to stretch the glide. Bloomfield told me that immediately before his last space flight, in the spring of 2002, his crew and a Mission Control team had gone through a full-dress simulation during which the orbiter had lost all three engines by Mach 21.7 (less than 100 mph from the decision speed). Confident in his ability to fly the more difficult Canadian arrival, Bloomfield, from the cockpit of the simulator, radioed, "We're going high-energy into Gander."

Mission Control answered, "Negative," and called for Shannon instead.

Bloomfield looked over at his right-seat pilot and said, "I think we oughta go to Gander. What do you think?"

"Yeah."

Bloomfield radioed back: "No, we think we oughta go to Gander."

Mission Control was emphatic. "Negative. We see you having enough energy to make Shannon."

As commander, Bloomfield had formal authority for the decision, but Mission Control, with its expert teams and wealth of data, was expressing a strong opinion, so he acquiesced. Acquiescence is standard in such cases, and usually it works out for the best. Bloomfield had enormous respect for the expertise and competence of Mission Control. He was also well aware of errors he had made in the past, despite superior advice or instructions from the flight controllers. This time, however, it turned out that two of the flight controllers had not communicated correctly with each other, and that the judgment of Mission Control therefore was wrong. Lacking the energy to reach Shannon, the simulator went into the ocean well short of the airport. The incident caused a disturbance inside the Johnson Space Center, particularly because of the longstanding struggle for the possession of data (and ultimately control) between the pilots in flight and the engineers at their consoles. Nevertheless, the two groups worked together, hammered out the problems, and the next day flew the same simulator profile successfully. But that was not the point of Bloomfield's story. Rather, it was that these calls are hard to make, and that mistakes—whether his or the controllers'—may become obvious only after it is too late.

For all its realism, the simulator cannot duplicate the gravity load of the climb, or the lack of it at the top. The transition to weightlessness is abrupt, and all the more dramatic because it occurs at the end of the 3 G acceleration: when the main engines cut off, the crew gets the impression of going over an edge and suddenly dropping into a free fall. That impression is completely accurate. In fact the term zero gravity (0 G), which is loosely used to describe the orbital environment, refers to physical acceleration, and does not mean that Earth's gravitational pull has somehow gone away. Far from it: the diminution of gravitational pull that comes with distance is small at these low-orbit altitudes (perhaps 200 miles above the surface), and the shuttle is indeed now falling—about like a stone dropped off a cliff. The fall does not, of course, diminish the shuttle's mass (if it bumps the Space Station, it does so with tremendous force), but it does make the vehicle and everything inside it very nearly weightless. The orbital part of the trick is that though the shuttle is dropping like a stone, it is also progressing across Earth's surface so fast (17,500 mph) that its path matches (roughly) the curvature of the globe. In other words, as it plummets toward the ground, the ground keeps getting out of its way. Like the orbits of all other satellites, and of the Space Station, and of the Moon as well, its flight is nothing but an unrestricted free fall around and around the world.

To help the astronauts adapt to weightlessness, the quarters are designed with a conventional floor-down orientation. This isn't quite so obvious as it might seem, since the shuttle flies inverted in orbit. "Down" therefore is toward outer space—and the view from the cockpit windows just happens to be of Earth sliding by from behind and overhead. The crews are encouraged to live and work with their heads "up" nonetheless. It is even recommended that they use the ladder while passing through the hatch between the two levels, and that they "descend" from the cockpit to the mid-deck feet first. Those sorts of cautions rarely prevail against the temptations of weightlessness. After Bloomfield's last flight one of his crew commented that they had all been swimming around "like eels in a can." Or like superhumans, as the case may be. It's true that there are frustrations: if you try to throw a switch without first anchoring your body, the switch will throw you. On the other hand, once you are anchored, you can shift multi-ton masses with your fingertips. You can also fly without wings, perform unlimited flips, or simply float for a while, resting in midair. Weightlessness is bad for the bones, but good for the soul. I asked Bloomfield how it had felt to experience gravity again. He said he remembered the first time, after coming to a stop on the runway in Florida, when he picked up a small plastic checklist in the cockpit and thought, "Man, this is so heavy!" He looked at me and said, "Gravity sucks."

And orbital flight clearly does not. The ride is smooth. When the cabin ventilation is turned off, as it must be once a day to exchange the carbon dioxide scrubbers, the silence is absolute. The smell inside the shuttle is distinctly metallic, unless someone has just come in from a spacewalk, after which the quarters are permeated for a while with "the smell of space," a pungent burned odor that some compare to that of seared meat, and that Bloomfield describes as closer to the smell of a torch on steel. The dominant sensation, other than weightlessness, is of the speed across the ground. Bloomfield said, "From California to New York in ten minutes, around the world once in ninety minutes—I mean, we're moving." He told me that he took to loitering in the cockpit at the end of the workdays, just for the view. By floating forward above the instrument panel and wrapping his legs around one of the pilot seats, he could position his face so close to the front windshield that the structure of the shuttle would seem to disappear.

The view from there was etched into his memory as a continuous loop. In brief, he said: It's night and you're coming up on California, with that clearly defined coastline, and you can see all the lights all the way from Tijuana to San Francisco, and then it's behind you, and you spot Las Vegas and its neon-lit Strip, which you barely have time to identify before you move across the Rockies, with their helter-skelter of towns, and then across the Plains, with its monotony of look-alike wheels and spokes of light, until you come to Chicago and its lakefront, from which point you can see past Detroit and Cleveland all the way to New York. These are big cities, you think. And because you grew up on a farm in Michigan, played football there in high school, and still know it like a home, you pick out Ann Arbor and Flint, and the place where I-75 joins U.S. Highway 23, and you get down to within a couple of miles of your house before zip, you're gone. Zip goes Cleveland, and zip New York, and then you're out over the Atlantic beyond Maine, looking back down the eastern seaboard all the way past Washington, DC. Ten minutes later you come up on Europe, and you hardly have time to think that London is a sprawl, France is an orderly display, the Alps are the Rockies again, and Italy is indeed a boot. Over Sicily you peer down into Etna's crater, into the glow of molten rock on Earth's inside, and then you are crossing Africa, where the few lights you see are not yellow but orange, like open flames. Past the Equator and beyond Madagascar you come to a zone of gray between the blackness of the night and the bright blue of the day. At the center of that zone is a narrow pink slice, which is the atmospheric dawn as seen from above. Daylight is for the oceans—first the Indian and then the Pacific, which is very, very large. Atolls appear with coral reefs and turquoise lagoons, but mostly what you see is cloud and open water. Then the pink slice of sunset passes below, and the night, and soon afterward you come again to California, though at another point on the coast, because ninety minutes have passed since you were last here, and during that time the world has revolved beneath you.

Ultimately the shuttle must return to Earth and land. The problem then is what to do with the vast amount of physical energy that has been invested in it—almost all the calories once contained in the nearly four million pounds of rocket fuel that was used to shove the shuttle into orbit. Some of that energy now resides in the vehicle's altitude, but most resides in its speed. The re-entry is a descent to a landing, yes, but primarily it is a giant deceleration, during which atmospheric resistance is used to convert velocity into heat, and to slow the shuttle by roughly 17,000 mph, so that it finally passes overhead the runway in Florida at airline speeds, and circles down to touch the ground at a well tamed 224 mph or less. Once the shuttle is on the runway, the drag chute and brakes take care of the rest.

The re-entry is a one-way ride that cannot be stopped once it has begun. The opening move occurs while the shuttle is still proceeding tail first and inverted, halfway around the world from the runway, high above the Indian Ocean. It is a simple thing, a brief burn by the twin orbital maneuvering rockets against the direction of flight, which slows the shuttle by perhaps 200 mph. That reduction is enough. The shuttle continues to free-fall as it has in orbit, but it now lacks the speed to match the curvature of Earth, so the ground no longer gets out of its way. By the time it reaches the start of the atmosphere, the "entry interface" at 400,000 feet, it has gently flipped itself around so that it is right-side up and pointed for Florida, but with its nose held 40 degrees higher than the angle of the descent path. The effect of this so-called angle of attack (which technically refers to the wings, not the nose) is to create drag, and to shield the shuttle's internal structures from the intense re-entry heat by cocking the vehicle up to greet the atmosphere with leading edges made of heat-resistant carbon-composite panels, and with 24,305 insulating surface tiles, each one unique, which are glued primarily to the vehicle's underside. To regulate the sink and drag (and to control the heating), the shuttle goes through a program of sweeping S-turns, banking as steeply as 80 degrees to one side and then the other, tilting its lift vector and digging into the atmosphere. The thinking is done by redundant computers, which use onboard inertial sensing systems to gauge the shuttle's position, altitude, descent rate, and speed. The flying is done by autopilot. The cockpit crews and mission controllers play the role of observers, albeit extremely interested ones who are ready to intervene should something go wrong. In a basic sense, therefore, the re-entry is a mirror image of the launch and climb, decompressed to forty-five minutes instead of eight, but with the added complication that it will finish with the need for a landing.

Bloomfield took me through it in simulation, the two of us sitting in the cockpit to watch while an experienced flight crew and full Mission Control team brought the shuttle in from the de-orbit burn to the touchdown, dealing with a complexity of cascading system failures. Of course, in reality the automation usually performs faultlessly, and the shuttle proceeds to Florida right on track, and down the center of the desired descent profile. Bloomfield expressed surprise at how well the magic had worked on his own flights. Because he had launched on high-inclination orbits to the Russian station *Mir* and the International Space Station, he had not flown a *Columbia*-style re-entry over the United States, but had descended across Central America instead. He said, "You look down over Central America, and you're so low that you can see the forests! You think, 'There's no way we're going to make it to Florida!' Then you cross the west coast of Florida, and you look inside, and you're still doing Mach 5, and you think, 'There's no way we're going to slow in time!' " But you do. Mach 5 is 3,500 mph. At that point the shuttle is at 117,000 feet, about 140 miles out. At Mach 2.5, or 1,650 mph, it is at 81,000 feet, about sixty miles out. At that point the crew activates the head-up displays, which project see-through flight guidance into the field of vision through the windshield. When the shuttle slows below the speed of sound, it shudders as the shock waves shift. By tradition if not necessity, the commander then takes over from the autopilot, and flies the rest of the arrival manually, using the control stick.

Bloomfield invited me to fly some simulated arrivals myself, and prompted me while I staggered around for a few landings—overhead the Kennedy Space Center at 30,000 feet with the runway and the coastal estuaries in sight below, banking left into a tight, plunging energy-management turn, rolling out onto final approach at 11,000 feet, following an extraordinarily steep, 18-degree glide slope at 345 mph, speed brakes on, pitching up through a "pre-flare" at 2,000 feet to flatten the descent, landing gear out at 300 feet, touching down on the main wheels with some skips and bumps, then drag chute out, nose gear gently down, and brakes on. My efforts were crude, and greatly assisted by Bloomfield, but they gave me an impression of the shuttle as a solid, beautifully balanced flying machine that in thick air, at the end, is responsive and not difficult to handle—if everything goes just right. Bloomfield agreed. Moreover, years have passed in which everything did go just right—leaving the pilots to work on the finesse of their touchdowns, whether they were two knots fast, or 100 feet long. Bloomfield said, "When you come back and you land, the engineers will pull out their charts and they'll say things like 'The boundary layer tripped on the left wing before the right one. Did you feel anything?' And the answer is always 'Well . . . no. It was an incredibly smooth ride all the way down.'" But then, on the morning of February 1, something went really wrong—something too radical for simulation, that offered the pilots no chance to fly—and the *Columbia* lay scattered for 300 miles across the ground.

The foam did it. That much was suspected from the start, and all the evidence converged on it as the CAIB's investigation proceeded through the months that followed. The foam was dense and dry; it was the brownish-orange coating applied to the outside of the shuttle's large external tank to insulate the extreme cold of the rocket fuels inside from the warmth and moisture of the air. Eighty-two seconds after liftoff, as the *Columbia* was accelerating through 1,500 mph, a piece of that foam—about nineteen inches long by eleven inches wide, weighing about 1.7 pounds—broke off from the external tank and collided with the left wing at about 545 mph. Cameras near the launch site recorded the event—though the images when viewed the following day provided insufficient detail to know the exact impact point, or the consequences. The CAIB's investigation ultimately found that a gaping hole about ten inches across had been punched into the wing's leading edge, and that six-teen days later the hole allowed the hot gases of the re-entry to penetrate the wing and consume it from the inside. Through enormous effort this would be discovered and verified beyond doubt. It was important nonetheless to explore the alternatives. In an effort closely supervised by the CAIB, groups of NASA engineers created several thousand flow charts, one for each scenario that could conceivably have led to the re-entry breakup. The thinking was rigorous. For a scenario to be "closed," meaning set aside, absolute proof had to be found (usually physical or mathematical) that this particular explanation did not apply: there was no cockpit fire, no flight-control malfunction, no act of terrorism or sabotage that had taken the shuttle down. Unexpected vulnerabilities were found during this process, and even after the investigation was formally concluded, in late August, more than a hundred scenarios remained technically open, because they could not positively be closed. For lack of evidence to the contrary, for instance, neither bird strikes nor micrometeorite impacts could be completely ruled out.

But for all their willingness to explore less likely alternatives, many of NASA's managers remained stubbornly closed-minded on the subject of foam. From the earliest telemetric data it was known that intense heat inside the left wing had destroyed the *Columbia,* and that such heat could have gotten there only through a hole. The connection between the hole and the foam strike was loosely circumstantial at first, but it required serious consideration nonetheless. NASA balked at going down that road. Its reasons were not rational and scientific but, rather, complex and cultural, and they turned out to be closely related to the errors that had led to the accident in the first place: simply put, it had become a matter of faith within NASA that foam strikes—which were a known problem—could not cause mortal damage to the shuttle. Sean O'Keefe, who was badly advised by his NASA lieutenants, made unwise public statements deriding the "foamologists"; and even Ron Dittemore, NASA's technically expert shuttle program manager, joined in with categorical denials.

At the CAIB, Gehman, who was not unsympathetic to NASA, watched these reactions with growing skepticism and a sense of déjà vu. Over his years in the Navy, and as a result of the *Cole* inquiry, he had become something of a student of large organizations under stress. To me he said, "It has been scorched into my mind that bureaucracies will do anything to defend themselves. It's not evil—it's just a natural reaction of bureaucracies, and since NASA is a bureaucracy, I expect the same out of them. As we go through the investigation, I've been looking for signs where the system is trying to defend itself." Of those signs the most obvious was this display of blind faith by an organization dependent on its engineering cool; NASA, in its absolute certainty, was unintentionally signaling the very problem that it had. Gehman had seen such certainty proved wrong too many times, and he told me that he was not about to get "rolled by the system," as he had been rolled before. He said, "Now when I hear NASA telling me things like 'Gotta be true!' or 'We know this to be true!' all my alarm bells go off. . . . Without hurting anybody's feelings, or squashing people's egos, we're having to say, 'We're sorry, but we're not accepting that answer.' "

That was the form that the physical investigation took on, with hundreds of NASA engineers and technicians doing most of the detailed work, and the CAIB watching closely and increasingly stepping in. Despite what Gehman said, it was inevitable that feelings got hurt and egos squashed—and indeed that serious damage to people's lives and careers was inflicted. At the NASA facilities dedicated to shuttle operations (Alabama for rockets, Florida for launch and landing, Texas for management and mission control) the CAIB investigators were seen as invaders of sorts, unwelcome strangers arriving to pass judgment on people's good-faith efforts. On the ground level, where the detailed analysis was being done, there was active resistance at first, with some NASA engineers openly refusing to cooperate, or to allow access to records and technical documents that had not been pre-approved for release. Gehman had to intervene. One of the toughest and most experienced of the CAIB investigators later told me he had a gut sense that NASA continued to hide relevant information, and that it does so to this day. But cooperation between the two groups gradually improved as friendships were made, and the intellectual challenges posed by the inquiry began to predominate over fears about what had happened or what might follow. As so often occurs, it was on an informal basis that information flowed best, and that much of the truth was discovered.

Board member Steven Wallace described the investigation not as a linear path but as a picture that gradually filled in. Or as a jigsaw puzzle. The search for debris began the first day, and soon swelled to include more than 25,000 people, at a cost of well over $300 million. NASA received 1,459 debris reports, including some from nearly every state in the union, and also from Canada, Jamaica, and the Bahamas. Discounting the geographic extremes, there was still a lot to follow up on. Though the amateur videos showed pieces separating from the shuttle along the entire path over the United States, and though search parties backtracked all the way to the Pacific coast in the hope of finding evidence of the breakup's triggering mechanism, the westernmost piece found on the ground was a left-wing tile that landed near a town called Littlefield, in the Texas Panhandle. Not surprisingly, the bulk of the wreckage lay under the main breakup, from south of Dallas eastward across the rugged, snake-infested brushland of East Texas and into Louisiana; and that is where most of the search took place. The best work was done on foot, by tough and dedicated crews who walked in tight lines across several thousand square miles. Their effort became something of a close sampling of the American landscape, turning up all sorts of odds and ends, including a few apparent murder victims, plenty of junked cars, and the occasional clandestine meth lab. More to the point, it also turned up crew remains and more than 84,000 pieces of the *Columbia*, which, at 84,900 pounds, accounted for 38 percent of the vehicle's dry weight. Certain pieces that had splashed into the murky waters of lakes and reservoirs were never found. It was presumed that most if not all the remaining pieces had been vaporized by the heat of re-entry, either before or after the breakup.

Some of the shuttle's contents survived intact. For instance, a vacuum cleaner still worked, as did some computers and printers and a Medtronic Tono-Pen, used to measure ocular pressure. A group of worms from one of the science experiments not only survived but continued to multiply. Most of the debris, however, was a twisted mess. The recovered pieces were meticulously plotted and tagged, and transported to a hangar at the Kennedy Space Center, where the wing remnants were laid out in correct position on the floor, and what had been found of the left wing's reinforced carbon-carbon (RCC) leading edge was reconstructed in a transparent Plexiglas mold—though with large gaps where pieces were missing. The hangar was a quiet, poignant, intensely focused place, with many of the same NASA technicians who had prepared the *Columbia* for flight now involved in the sad task of handling its ruins. The assembly and analysis went on through the spring. One of the principal CAIB agents there was an affable Air Force pilot named Patrick Goodman, an experienced accident investigator who had made both friends and enemies at NASA for the directness of his approach. When I first met him, outside the hangar on a typically warm and sunny Florida day, he explained some of the details that I had just seen on the inside—heat-eroded tiles, burned skin and structure, and aluminum slag that had emerged in molten form from inside the left wing, and had been deposited onto the aft rocket pods. The evidence was complicated because it resulted from combinations of heat, physical forces, and wildly varying airflows that had occurred before, during, and after the main-body breakup, but for Goodman it was beginning to read like a map. He had faith. He said, "We know what we have on the ground. It's the truth. The debris is the truth, if we can only figure out what it's saying. It's not a theoretical model. It exists." Equally important was the debris that did not exist, most significantly large parts of the left wing, including the lower part of a section of the RCC leading edge, a point known as Panel Eight, which was approximately where the launch cameras showed that the foam had hit. Goodman said, "We look at what we don't have. What we do have. What's on what we have. We start from there, and try to work backwards up the timeline, always trying to see the previous significant event." He called this "looking uphill." It was like a movie run in reverse, with the found pieces springing off the ground and flying upward to a point of reassembly above Dallas, and then the *Columbia*, looking nearly whole, flying tail-first toward California, picking up the Littlefield tile as it goes, and then higher again, through entry interface over the Pacific, through orbits flown in reverse, inverted but nose first, and then back down toward Earth, picking up the external tank and the solid rocket boosters during the descent, and settling tail-first with rockets roaring, until just before a vertical touchdown a spray of pulverized foam appears below, pulls together at the left-wing leading edge, and rises to lodge itself firmly on the side of the external tank.

The foam did it.

There was plenty of other evidence, too. After the accident the Air Force dug up routine radar surveillance tapes that upon close inspection showed a small object floating alongside the *Columbia* on the second day of its mission. The object slowly drifted away and disappeared from view. Subsequent testing of radar profiles and ballistic coefficients for a multitude of objects found a match for only one—a fragment of RCC panel of at least 140 square inches. The match never quite passed muster as proof, but investigators presumed that the object was a piece of the leading edge, that it had been shoved into the inside of the wing by the impact of the foam, and that during maneuvering in orbit it had floated free. The picture by now was rapidly filling in.

But the best evidence was numerical. It so happened that because the *Columbia* was the first of the operational shuttles, it was equipped with hundreds of additional engineering sensors that fed into an onboard data-collection device, a box known as a modular auxiliary data system, or MADS recorder, that was normally used for post-flight analysis of the vehicle's performance. During the initial debris search this box was not found, but such was its potential importance that after careful calculation of its likely ballistic path, another search was mounted, and on March 19 it was discovered—lying in full view on ground that had been gone over before. The really surprising thing was its condition. Though the recorder was not designed to be crashproof, and used Mylar tape that was vulnerable to heat, it had survived the breakup and fall completely intact, as had the data that it contained, the most interesting of which pertained to heat rises and sequential sensor failures inside the left wing. When combined with the telemetric data that already existed, and with calculations of the size and location of the sort of hole that might have been punched through the leading edge by the foam, the new data allowed for a good fit with computational models of the theoretical airflow and heat propagation inside the left wing, and it steered the investigation to an inevitable conclusion that the breach must have been in the RCC at Panel Eight.

By early summer the picture was clear. Though strictly speaking the case was circumstantial, the evidence against the foam was so persuasive that there remained no reasonable doubt about the physical cause of the accident. As a result, Gehman gave serious consideration to NASA's request to call off a planned test of the launch incident, during which a piece of foam would be carefully fired at a fully rigged RCC Panel Eight. NASA's argument against the test had some merit: the leading-edge panels (forty-four per shuttle) are custom-made, $700,000 components, each one different from the others, and the testing would require the use of the last spare Panel Eight in the entire fleet. NASA said that it couldn't afford the waste, and Gehman was inclined to agree, precisely because he felt that breaking the panel would prove nothing that hadn't already been amply proved. By a twist of fate it was the sole NASA member of the CAIB, the quiet, cerebral, earnestly scientific Scott Hubbard, who insisted that the test proceed. Hubbard was one of the original seven board members. At the time of the accident he had just become the director of NASA's Ames Research Center, in California. Months later now, in the wake of Gehman's rebellion, and with the CAIB aggressively moving beyond the physical causes and into the organizational ones, he found himself in the tricky position of collaborating with a group that many of his own people at NASA saw as the enemy. Hubbard, however, had an almost childlike belief in doing the right thing, and having been given this unfortunate job, he was determined to see it through correctly. Owing to the closeness of his ties to NASA, he understood an aspect of the situation that others might have overlooked: despite overwhelming evidence to the contrary, many people at NASA continued stubbornly to believe that the foam strike on launch could not have caused the *Columbia*'s destruction. Hubbard argued that if NASA was to have any chance of self-reform, these people would have to be confronted with reality, not in abstraction but in the most tangible way possible. Gehman found the argument convincing, and so the foam shot proceeded.

The work was done in San Antonio, using a compressed-nitrogen gun with a thirty-five-foot barrel, normally used to fire dead chickens—real and artificial—against aircraft structures in bird-strike certification tests. NASA approached the test kicking and screaming all the way, insisting, for instance, that the shot be used primarily to validate an earlier debris-strike model (the so-called Crater model of strikes against the underside tiles) that had been used for decision-making during the flight, and was now known to be irrelevant. Indeed, it was because of NASA obstructionism—and specifically the illogical insistence by some of the NASA rocket engineers that the chunk of foam that had hit the wing was significantly smaller (and therefore lighter) than the video and film record showed it to be—that the CAIB and Scott Hubbard finally took direct control of the testing. There was in fact a series of foam shots, increasingly realistic according to the evolving analysis of the actual strike, that raised the stakes from a glancing blow against the underside tiles to steeper-angle hits directly against leading-edge panels. The second to last shot was a 22-degree hit against the bottom of Panel Six: it produced some cracks and other damage deemed too small to explain the shuttle's loss. Afterward there was some smugness at NASA, and even Sean O'Keefe, who again was badly advised, weighed in on the matter, belittling the damage. But the shot against Panel Six was not yet the real thing. That was saved for the precious Panel Eight, in a test that was painstakingly designed to duplicate (conservatively) the actual impact against the *Columbia*'s left wing, assuming a rotational "clocking angle" 30 degrees off vertical for the piece of foam. Among the engineers who gathered to watch were many of those still living in denial. The gun fired, and the foam hit the panel at a 25-degree relative angle at about 500 mph. Immediately afterward an audible gasp went through the crowd. The foam had knocked a hole in the RCC large enough to allow people to put their heads through. Hubbard told me that some of the NASA people were close to tears. Gehman had stayed away in order to avoid the appearance of gloating. He could not keep the satisfaction out of his voice, however, when later he said to me, "Their whole house of cards came falling down."

NASA's house was by then what this investigation was really all about. The CAIB discovered that on the morning of January 17, the day after the launch, the low-level engineers at the Kennedy Space Center whose job was to review the launch videos and film were immediately concerned by the size and speed of the foam that had struck the shuttle. As expected of them, they compiled the imagery and disseminated it by e-mail to various shuttle engineers and managers—most significantly those in charge of the shuttle program at the Johnson Space Center. Realizing that their blurred or otherwise inadequate pictures showed nothing of the damage that might have been inflicted, and anticipating the need for such information by others, the engineers at Kennedy then went outside normal channels and on their own initiative approached the

Department of Defense with a request that secret military satellites or ground-based high-resolution cameras be used to photograph the shuttle in orbit. After a delay of several days for the back-channel request to get through, the Air Force proved glad to oblige, and made the first moves to honor the request. Such images would probably have shown a large hole in the left wing—but they were never taken.

When news of the foam strike arrived in Houston, it did not seem to be crucially important. Though foam was not supposed to shed from the external tank, and the shuttle was not designed to withstand its impacts, falling foam had plagued the shuttle from the start, and indeed had caused damage on most missions. The falling foam was usually popcorn sized, too small to cause more than superficial dents in the thermal protection tiles. The CAIB, however, discovered a history of more-serious cases. For example, in 1988 the shuttle *Atlantis* took a heavy hit, seen by the launch cameras eighty-five seconds into the climb, nearly the same point at which the *Columbia* strike occurred. On the second day of the *Atlantis* flight Houston asked the crew to inspect the vehicle's underside with a video camera on a robotic arm (which the *Columbia* did not have). The commander, Robert "Hoot" Gibson, told the CAIB that the belly looked as if it had been blasted with shotgun fire. The *Atlantis* returned safely anyway, but afterward was found to have lost an entire tile, exposing its bare metal belly to the re-entry heat. It was lucky that the damage had happened in a place where a heavy aluminum plate covered the skin, Gibson said, because otherwise the belly might have been burned through.

Nonetheless, over the years foam strikes had come to be seen within NASA as an "in-family" problem, so familiar that even the most serious episodes seemed unthreatening and mundane. Douglas Osheroff, a normally good-humored Stanford physicist and Nobel laureate who joined the CAIB late, went around for months in a state of incredulity and dismay at what he was learning about NASA's operational logic. He told me that the shuttle managers acted as if they thought the frequency of the foam strikes had somehow reduced the danger that the impacts posed. His point was not that the managers really believed this but that after more than a hundred successful flights they had come blithely to accept the risk. He said, "The excitement that only exists when there is danger was kind of gone—even though the danger was not gone." And frankly, organizational and bureaucratic concerns weighed more heavily on the managers' minds. The most pressing of those concerns were the new performance goals imposed by Sean O'Keefe, and a tight sequence of flights leading up to a drop-dead date of February 19, 2004, for the completion of the International Space Station's "core." O'Keefe had made it clear that meeting this deadline was a test, and that the very future of NASA's human space-flight program was on the line.

From Osheroff's scientific perspective, deadlines based on completion of the International Space Station were inherently absurd. To me he said, "And what would the next goal be after that? Maybe we should bring our pets up there! 'I wonder how a Saint Bernard urinates in zero gravity!' NASA sold the International Space Station to Congress as a great science center—but most scientists just don't agree with that. We're thirty years from being able to go to Mars. Meanwhile, the only reason to have man in space is to study man in space. You can do that stuff—okay—and there are also some biology experiments that are kind of fun. I think we are learning things. But I would question any statement that you can come up with better drugs in orbit than you can on the ground, or that sort of thing. The truth is, the International Space Station has become a huge liability for NASA"—expensive to build, expensive to fly, expensive to resupply. "Now members of Congress are talking about letting its orbit decay—just letting it fall into the ocean. And it does turn out that orbital decay is a very good thing, because it means that near space is a self-cleaning place. I mean, garbage does not stay up there forever."

In other words, completion of the Space Station could provide a measure of NASA's performance only in the most immediate and superficial manner, and it was therefore an inherently poor reason for shuttle managers to be ignoring the foam strikes and proceeding at full speed. It was here that you could see the limitations of leadership without vision, and the consequences of putting an executive like O'Keefe in charge of an organization that needed more than mere discipline. This, however, was hardly an argument that the managers could use, or even in private allow themselves to articulate. If the Space Station was unimportant—and perhaps even a mistake—then one had to question the reason for the shuttle's existence in the first place. Like O'Keefe and the astronauts and NASA itself, the managers were trapped by a circular space policy thirty years in the making, and they had no choice but to strive to meet the timelines directly ahead. As a result, after the most recent *Atlantis* launch, in October of 2002, during which a chunk of foam from a particularly troublesome part of the external tank, known as the "bipod ramp," had dented one of the solid rocket boosters, shuttle managers formally decided during the post-flight review not to classify the incident as an "in-flight anomaly." This was the first time that a serious bipod-ramp incident had escaped such a classification. The decision allowed the following two launches to proceed on schedule. The second of those launches was the *Columbia*'s, on January 16.

The videos of the foam strike reached Houston the next day, January 17.

They made it clear that again the offending material had come from the area of the bipod ramp, that this time the foam was larger than ever before, that the impact had occurred later in the climb (meaning at higher speed), and that the wing had been hit, though exactly where was not clear. The astronauts were happily in orbit now, and had apparently not felt the impact, or been able to distinguish it from the heavy vibrations of the solid rocket boosters. In other words, they were unaware of any trouble. Responsibility for disposing of the incident lay with engineers on the ground, and specifically with the Mission Management Team, or MMT, whose purpose was to make decisions about the problems and unscripted events that inevitably arose during any flight. The MMT was a high-level group. In the Houston hierarchy it operated above the flight controllers in the Mission Control room, and just below the shuttle program manager, Ron Dittemore. Dittemore was traveling at the time, and has since retired. The MMT meetings were chaired by his protégé, the once rising Linda Ham, who has come to embody NASA's arrogance and insularity in many observers' minds. Ham is the same hard-charging manager who, with a colleague, later had to be forcefully separated from the CAIB's investigation. Within the strangely neutered engineering world of the Johnson Space Center, she was an intimidating figure, a youngish, attractive woman given to wearing revealing clothes, yet also known for a tough and domineering management style. Among the lower ranks she had a reputation for brooking no nonsense and being a little hard to talk to. She was not smooth. She was a woman struggling upward in a man's world. She was said to have a difficult personality.

As the head of the MMT, Ham responded to news of the foam strike as if it were just another item to be efficiently handled and then checked off the list: a water leak in the science lab, a radio communication failure, a foam strike on the left wing, okay, no safety-of-flight issues here—right? What's next? There was a trace of vanity in the way she ran her shows. She seemed to revel in her own briskness, in her knowledge of the shuttle systems, in her use of acronyms and the strange, stilted syntax of aerospace engineers. She was decisive, and very sure of her sense for what was important and what was not. Her style got the best of her on day six of the mission, January 21, when at a recorded MMT meeting she spoke just a few words too many, much to her later regret.

It was at the end of a report given by a mid-ranking engineer named Don McCormack, who summarized the progress of an ad hoc engineering group, called the Debris Assessment Team, that had been formed at a still lower level to analyze the foam strike. The analysis was being done primarily by Boeing engineers, who had dusted off the soon to be notorious Crater model, primarily to predict damage to the underwing tile. McCormack reported that little was yet resolved, that the quality of the Crater as a predictor was being judged against the known damage on earlier flights, and that some work was being done to explore the options should the analysis conclude that the *Columbia* had been badly wounded. After a brief exchange Ham cut him short, saying, "And I'm really . . . I don't think there is much we can do, so it's not really a factor during the flight, since there is not much we can do about it." She was making assumptions, of course, and they were later proved to be completely wrong, but primarily she was just being efficient, and moving the meeting along. After the accident, when the transcript and audiotapes emerged, those words were taken out of context, and used to portray Ham as a villainous and almost inhumanly callous person, which she certainly was not. In fact, she was married to an astronaut, and was as concerned as anyone about the safety of the shuttle crews. This was a dangerous business, and she knew it all too well. But like her boss, Ron Dittemore, with whom she discussed the *Columbia* foam strike several times, she was so immersed in the closed world of shuttle management that she simply did not elevate the event—this "in-family" thing—to the level of concerns requiring action. She was intellectually arrogant, perhaps, and as a manager she failed abysmally. But neither she nor the others of her rank had the slightest suspicion that the *Columbia* might actually go down.

THE FRUSTRATION IS that some people on lower levels were actively worried about that possibility, and they understood clearly that not enough was known about the effects of the foam strike on the wing, but they expressed their concerns mostly to one another, and for good reason, because on the few occasions when they tried to alert the decision-makers, NASA's management system overwhelmed them and allowed none of them to be heard. The question now, of course, is why.

The CAIB's search for answers began long before the technical details were resolved, and it ultimately involved hundreds of interviews and 50,000 pages of transcripts. The manner in which those interviews were conducted became a contentious issue, and it was arguably Gehman's biggest mistake. As a military man, advised by military men on the board, he decided to conduct the interviews according to a military model of safety probes, in which individual fault is not formally assigned, and the interviews themselves are "privileged," meaning forever sealed off from public view. It was understood that identities and deeds would not be protected from view, only individual testimonies to the CAIB, but serious critics cried foul nonetheless, and pointed out correctly that Gehman was using loopholes to escape sunshine laws that otherwise would have applied. Gehman believed that treating the testimony as privileged was necessary to encourage witnesses to talk, and to get to the bottom of the story, but the long-term effect of the investigation will be diminished as a result (for instance, by lack of access to the raw material by outside analysts), and there was widespread consensus among the experienced (largely civilian) investigators actually conducting the interviews that the promise of privacy was having little effect on what people were willing to say. These were not criminals they were talking to, or careful lawyers. For the most part they were sincere engineering types who were concerned about what had gone wrong, and would have been willing even without privacy to speak their minds. The truth, in other words, would have come out even in the brightest of sunshine.

The story that emerged was a sad and unnecessary one, involving arrogance, insularity, and bad luck allowed to run unchecked. On the seventh day of the flight, January 22, just as the Air Force began to move on the Kennedy engineers' back-channel request for photographs, Linda Ham heard to her surprise that this approach (which according to front-channel procedures would have required her approval) had been made. She immediately telephoned other high-level managers in Houston to see if any of them wanted to issue a formal "requirement" for imagery, and when they informed her that they did not, rather than exploring the question with the Kennedy engineers she simply terminated their request with the Department of Defense. This appears to have been a purely bureaucratic reaction. A NASA liaison officer then e-mailed an apology to Air Force personnel, assuring them that the shuttle was in "excellent shape," and explaining that a foam strike was "something that has happened before and is not considered to be a major problem." The officer continued, "The one problem that this has identified is the need for some additional coordination within NASA to assure that when a request is made it is done through the official channels." Months later one of the CAIB investigators who had followed this trail was still seething with anger at what had occurred. He said, "Because the problem was not identified in the traditional way—'Houston, we have a problem!'—well, then, 'Houston, we don't have a problem!' Because Houston didn't identify the problem."

But another part of Houston was doing just that. Unbeknownst to Ham and the shuttle management, the low-level engineers of the Debris Assessment Team had concluded that the launch films were not clear enough to indicate where the foam had hit, and particularly whether it had hit the underside tile or a leading-edge RCC panel. Rather than trying to run their calculations in the blind, they had decided that they should do the simple thing and have someone take a look for damage. They had already e-mailed one query to the

engineering department, about the possibility of getting the astronauts themselves to take a short spacewalk and inspect the wing. It later turned out that
this would have been safe and easy to do. That e-mail, however, was never
answered. This time the Debris Assessment engineers decided on a still simpler
solution—to ask the Department of Defense to take some high-resolution pictures. Ignorant of the fact that the Kennedy group had already made such a
request, and that it had just been peevishly canceled, they sent out two requests
of their own, directed, appropriately, to Ron Dittemore and Linda Ham, but
through channels that were a little off-center, and happened to fail. Those
channels were ones they had used in their regular work as engineers, outside
the formal shuttle-management structure. By unfortunate circumstance, the
request that came closest to getting through was intercepted by a mid-level
employee (the assistant to an intended recipient, who was on vacation) who
responded by informing the Debris Assessment engineers, more or less correctly, that Linda Ham had decided against Air Force imagery.

The confusion was now total, yet also nearly invisible—and within the sup-
pressive culture of the human spaceflight program, it had very little chance of
making itself known. At the top of the tangle, neither Ron Dittemore nor Linda
Ham ever learned that the Debris Assessment Team wanted pictures; at the
bottom, the Debris Assessment engineers heard the "no" without suspecting
that it was not an answer to their request. They were told to go back to the
Crater model and numerical analysis, and as earnest, hardworking engineers
(hardly rebels, these), they dutifully complied, all the while regretting the blind
assumptions that they would have to make. Given the obvious potential for a
catastrophe, one might expect that they would have gone directly to Linda
Ham, on foot if necessary, to make the argument in person for a spacewalk or
high-resolution photos. However, such were the constraints within the Johnson Space Center that they never dared. They later said that had they made a
fuss about the shuttle, they might have been singled out for ridicule. They
feared for their standing, and their careers.

The CAIB investigator who asked the engineers what conclusion they had
drawn at the time from management's refusal later said to me, "They all
thought, 'Well, none of us have a security clearance high enough to view any of
this imagery.' They talked about this openly among themselves, and they figured one of three things:

" 'One: The "no" means that management's already got photos, and the
damage isn't too bad. They can't show us the photos, because we don't have
the security clearance, and they can't tell us they have the photos, or tell us
the damage isn't bad, because that tells us how accurate the photos are—and

we don't have the security clearance. But wait a minute, if that's the case, then what're we doing here? Why are we doing the analysis? So no, that can't be right.

" 'Okay, then, two: They already took the photos, and the damage is so severe that there's no hope for recovery. Well . . . that can't be right either, because in that case, why are we doing the analysis?

" 'Okay, then, three: They took the photos. They can't tell us they took the photos, and the photos don't give us clear definition. So we need to do the analysis. That's gotta be it!' "

What the Debris Assessment engineers could not imagine is that no photos had been taken, or ever would be—and essentially for lack of curiosity by NASA's imperious, self-convinced managers. What those managers in turn could not imagine was that people in their own house might really be concerned. The communication gap had nothing to do with security clearances, and it was complete.

Gehman explained the underlying realities to me. He said, "They claim that the culture in Houston is a 'badgeless society,' meaning it doesn't matter what you have on your badge—you're concerned about shuttle safety together. Well, that's all nice, but the truth is that it does matter what badge you're wearing. Look, if you really do have an organization that has free communication and open doors and all that kind of stuff, it takes a special kind of management to make it work. And we just don't see that management here. Oh, they say all the right things. 'We have open doors and e-mails, and anybody who sees a problem can raise his hand, blow a whistle, and stop the whole process.' But then when you look at how it really works, it's an incestuous, hierarchical system, with invisible rankings and a very strict informal chain of command. They all know that. So even though they've got all the trappings of communication, you don't actually find communication. It's very complex. But if a person brings an issue up, what caste he's in makes all the difference. Now, again, NASA will deny this, but if you talk to people, if you really listen to people, all the time you hear 'Well, I was afraid to speak up.' Boy, it comes across loud and clear. You listen to the meetings: 'Anybody got anything to say?' There are thirty people in the room, and slam! There's nothing. We have plenty of witness statements saying, 'If I had spoken up, it would have been at the cost of my job.' And if you're in the engineering department, you're a nobody."

One of the CAIB investigators told me that he asked Linda Ham, "As a manager, how do you seek out dissenting opinions?"

According to him, she answered, "Well, when I hear about them . . ."

He interrupted. "Linda, by their very nature you may not hear about them."
"Well, when somebody comes forward and tells me about them."
"But Linda, what techniques do you use to get them?"
He told me she had no answer.

This was certainly not the sort of risk-versus-risk decision-making that Michael Bloomfield had in mind when he described the thinking behind his own shuttle flights.

AT 7:00 AM ON THE NINTH DAY, January 24, which was one week before the *Columbia*'s scheduled re-entry, the engineers from the Debris Assessment Team formally presented the results of their numerical analysis to Linda Ham's intermediary, Don McCormack. The room was so crowded with concerned observers that some people stood in the hall, peering in. The fundamental purpose of the meeting would have been better served had the engineers been able to project a photograph of a damaged wing onto the screen, but, tragically, that was not to be. Instead they projected a typically crude PowerPoint summary, based on the results from the Crater model, with which they attempted to explain a nuanced position: first, that if the tile had been damaged, it had probably endured well enough to allow the *Columbia* to come home; and second, that for lack of information, they had needed to make assumptions to reach that conclusion, and that troubling unknowns therefore limited the meaning of the results. The latter message seems to have been lost. Indeed, this particular PowerPoint presentation became a case study for Edward Tufte, the brilliant communications specialist from Yale, who in a subsequent booklet, "The Cognitive Style of PowerPoint," tore into it for its dampening effect on clear expression and thought. The CAIB later joined in, describing the widespread use of PowerPoint within NASA as one of the obstacles to internal communication, and criticizing the Debris Assessment presentation for mechanically underplaying the uncertainties that remained.

Had the uncertainties been more strongly expressed as the central factor in question, the need to inspect the wing by spacewalk or photograph might have become obvious even to the shuttle managers. Still, the Mission Management Team seemed unprepared to hear nuance. Fixated on potential tile damage as the relevant question, assuming without good evidence that the RCC panels were strong enough to withstand a foam strike, subtly skewing the discussion away from catastrophic burn-through and toward the potential effects on turnaround times on the ground and how that might affect the all-important launch schedule, the shuttle managers were convinced that they had the situation as they defined it firmly under control.

At a regularly scheduled MMT meeting later that morning McCormack summarized the PowerPoint presentation for Linda Ham. He said, "The analysis is not complete. There is one case yet that they wish to run, but kind of just jumping to the conclusion of all that, they do show that [there is], obviously, a potential for significant tile damage here, but thermal analysis does not indicate that there is potential for a burn-through. I mean, there could be localized heating damage. There is . . . obviously there is a lot of uncertainty in all this in terms of the size of the debris and where it hit and the angle of incidence."

Ham answered, "No burn-through means no catastrophic damage. And the localized heating damage would mean a tile replacement?"

"Right, it would mean possible impacts to turnaround repairs and that sort of thing, but we do not see any kind of safety-of-flight issue here yet in anything that we've looked at."

This was all too accurate in itself. Ham said, "And no safety of flight, no issue for this mission, nothing that we're going to do different. There may be a turnaround [delay]."

McCormack said, "Right. It could potentially [have] hit the RCC . . . We don't see any issue if it hit the RCC . . ."

The discussion returned to the tiles. Ham consulted with a tile specialist named Calvin Schomburg, who for days had been energetically making a case independent of the Debris Assessment analysis that a damaged tile would endure re-entry—and thereby adding, unintentionally, to the distractions and false assumptions of the management team. After a brief exchange Ham cut off further discussion with a quick summary for some people participating in the meeting by conference call, who were having trouble hearing the speakerphone. She said, "So, no safety-of-flight kind of issue. It's more of a turnaround issue similar to what we've had on other flights. That's it? All right, any questions on that?"

And there were not.

For reasons unexplained, when the official minutes of the meeting were written up and distributed (having been signed off on by Ham), all mention of the foam strike was omitted. This was days before the *Columbia*'s re-entry, and seems to indicate sheer lack of attention to this subject, rather than any sort of cover-up.

The truth is that Linda Ham was as much a victim of NASA as were *Columbia*'s astronauts, who were still doing their science experiments then, and free-falling in splendor around the planet. Her predicament had roots that went way back, nearly to the time of Ham's birth, and it involved not only the culture of the human space-flight program but also the White House, Congress, and NASA leadership over the past thirty years. Gehman understood this fully, and as the investigation drew to a close, he vowed to avoid merely going after the people who had been standing close to the accident when it occurred. The person standing closest was, of course, Linda Ham, and she will bear a burden for her mismanagement. But by the time spring turned to summer, and the CAIB moved its operation from Houston to Washington, DC, Gehman had taken to saying, "Complex systems fail in complex ways," and he was determined that the CAIB's report would document the full range of NASA's mistakes. It did, and in clean, frank prose, using linked sentences and no PowerPoint displays.

As the report was released, on August 26, Mars came closer to Earth than it had in 60,000 years. Gehman told me that he continued to believe in the importance of America's human space-flight effort, and even of the return of the shuttle to flight—at least until a replacement with a clearer mission can be built and put into service. It was a quiet day in Washington, with Congress in recess and the President on vacation. Aides were coming from Capitol Hill to pick up several hundred copies of the report and begin planning hearings for the fall. The White House was receiving the report too, though keeping a cautious distance, as had been expected; it was said that the President might read an executive summary. Down in Houston, board members were handing copies to the astronauts, the managers, and the families of the dead.

Gehman was dressed in a suit, as he had been at the start of all this, seven months before. It was up to him now to drive over to NASA headquarters, in the southwest corner of the city, and deliver the report personally to Sean O'Keefe. I went along for the ride, as did the board member Sheila Widnall, who was there to lend Gehman some moral support. The car was driven by a Navy officer in whites. At no point since the accident had anyone at NASA stepped forward to accept personal responsibility for contributing to this accident—not Linda Ham, not Ron Dittemore, and certainly not Sean O'Keefe. However, the report in Gehman's hands (248 pages, full color, well bound) made responsibility very clear. This was not going to be a social visit. Indeed, it turned out to be extraordinarily tense. Gehman and Widnall strode up the carpeted hallways in a phalanx of anxious, dark-suited NASA staffers, who swung open the doors in advance and followed close on their heels. O'Keefe's office suite was practically imperial in its expense and splendor. High officials stood in small, nervous groups, murmuring. After a short delay O'Keefe appeared—a tall, balding, gray-haired man with stooped shoulders. He shook hands and ushered Gehman and Widnall into the privacy of his inner office. Ten minutes later they emerged. There was a short ceremony for NASA cameras, during which O'Keefe thanked Gehman for his important contribution, and then it was time to leave. As we drove away, I asked Gehman how it had been in there with O'Keefe.

He said "Stiff. Very stiff."

We talked about the future. The report had made a series of recommendations for getting the shuttle back into flight, and beyond that for beginning NASA's long and necessary process of reform. I knew that Gehman, along with much of the board, had volunteered to Congress to return in a year, to peer in deeply again, and to try to judge if progress had been made. I asked him how genuine he thought such progress could be, and he managed somehow to express hope, though skeptically.

BY JANUARY 23, the *Columbia*'s eighth day in orbit, the crew had solved a couple of minor system problems, and after a half day off, during which no doubt some of the astronauts took the opportunity for some global sightseeing, they were proceeding on schedule with their laboratory duties, and were in good spirits and health. They had been told nothing of the foam strike. Down in Houston, the flight controllers at Mission Control were aware of it, and they knew that the previous day Linda Ham had canceled the request for Air Force photographs. Confident that the issue would be satisfactorily resolved by the shuttle managers, they decided nonetheless to inform the flight crew by e-mail—if only because certain reporters at the Florida launch site had heard of it, and might ask questions at an upcoming press conference, a Public Affairs Office, or PAO, event. The e-mail was written by one of the lead flight controllers, in the standard, overly upbeat style. It was addressed to the pilots, Rick Husband and William McCool.

Under the subject line "INFO: Possible PAO Event Question," it read,

> Rick and Willie,
>
> You guys are doing a fantastic job staying on the timeline and accomplishing great science. Keep up the good work and let us know if there is anything that we can do better from an MCC/POCC standpoint.
>
> There is one item that I would like to make you aware of for the upcoming PAO event. . . . This item is not even worth mentioning other

than wanting to make sure that you are not surprised by it in a question from a reporter.

The e-mail then briefly explained what the launch pictures had shown—a hit from the bipod-ramp foam. A video clip was attached. The e-mail concluded,

> Experts have reviewed the high speed photography and there is no concern for RCC or tile damage. We have seen this same phenomenon on several other flights and there is absolutely no concern for entry. That is all for now. It's a pleasure working with you every day.

The e-mail's content honestly reflected what was believed on the ground, though in a repackaged and highly simplified form. There was no mention of the inadequate quality of the pictures, of the large size of the foam, of the ongoing analysis, or of Linda Ham's decision against Air Force imagery. This was typical for Mission Control communications, a small example of a long-standing pattern of something like information-hoarding that was instinctive and a matter as much of style as of intent: the astronauts had been told of the strike, but almost as if they were children who didn't need to be involved in the grown-up conversation. Two days later, when Rick Husband answered the e-mail, he wrote, "Thanks a million!" and "Thanks for the great work!" and after making a little joke, that "Main Wing" could sound like a Chinese name, he signed off with an e-mail smile —:). He made no mention of the foam strike at all. And with that, as we now know, the crew's last chance for survival faded away.

Linda Ham was wrong. Had the hole in the leading edge been seen, actions could have been taken to try to save the astronauts' lives. The first would have been simply to buy some time. Assuming a starting point on the fifth day of the flight, NASA engineers subsequently calculated that by requiring the crew to rest and sleep, the mission could have been extended to a full month, to February 15. During that time the *Atlantis,* which was already being prepared for a scheduled March 1 launch, could have been processed more quickly by ground crews working around the clock, and made ready to go by February 10. If all had proceeded perfectly, there would have been a five-day window in which to blast off, join up with the *Columbia,* and transfer the stranded astronauts one by one to safety, by means of tethered spacewalks. Such a rescue would not have been easy, and it would have involved the possibility of another fatal foam strike and the loss of two shuttles instead of one; but in the risk-versus-risk world of space flight, veterans like Mike Bloomfield would immediately have volunteered, and NASA would have bet the farm.

The fallback would have been a desperate measure—a jury-rigged repair performed by the *Columbia* astronauts themselves. It would have required two spacewalkers to fill the hole with a combination of heavy tools and metal scraps scavenged from the crew compartment, and to supplement that mass with an ice bag shaped to the wing's leading edge. In theory, if much of the payload had been jettisoned, and luck was with the crew, such a repair might perhaps have endured a modified re-entry and allowed the astronauts to bail out at the standard 30,000 feet. The engineers who came up with this plan realized that in reality it would have been extremely dangerous, and might well have led to a high-speed burn-through and the loss of the crew. But anything would have been better than attempting a normal re-entry as it was actually flown.

The blessing, if one can be found, is that the astronauts remained unaware until nearly the end. A home video shot on board and found in the wreckage documented the relaxed mood in the cockpit as the shuttle descended through the entry interface at 400,000 feet, at 7:44:09 Houston time, northwest of Hawaii. The astronauts were drinking water in anticipation of gravity's redistributive effect on their bodies. The *Columbia* was flying at the standard 40-degree nose-up angle, with its wings level, and still doing nearly 17,000 mph; outside, though the air was ultrathin and dynamic pressures were very low, the aerodynamic surfaces were beginning to move in conjunction with the array of control jets, which were doing the main work of maintaining the shuttle's attitude, and would throughout the re-entry. The astronauts commented like sightseers as sheets of fiery plasma began to pass by the windows.

The pilot, McCool, said, "Do you see it over my shoulder now, Laurel?"

Sitting behind him, the mission specialist Laurel Clark said, "I was filming. It doesn't show up nearly as much as the back."

McCool said to the Israeli payload specialist, Ilan Ramon, "It's going pretty good now. Ilan, it's really neat—it's a bright orange-yellow out over the nose, all around the nose."

The commander, Husband, said, "Wait until you start seeing the swirl patterns out your left or right windows."

McCool said, "Wow."

Husband said, "Looks like a blast furnace."

A few seconds later they began to feel gravity. Husband said, "Let's see here . . . look at that."

McCool answered, "Yup, we're getting some Cs." As if it were unusual, he said, "I let go of the card, and it falls." Their instruments showed that they were experiencing one hundredth of a G. McCool looked out the window again. He said, "This is amazing. It's really getting, uh, fairly bright out there."

Husband said, "Yup. Yeah, you definitely don't want to be outside now."

The flight engineer, Kalpana Chawla, answered sardonically, "What—like we did before?" The crew laughed.

Outside, the situation was worse than they imagined. Normally, as a shuttle streaks through the upper atmosphere it heats the air immediately around it to temperatures as high as 10,000°, but largely because of the boundary layer—a sort of air cushion created by the leading edges—the actual surface temperatures are significantly lower, generally around 3,000°, which the vehicle is designed to withstand, if barely. The hole in the *Columbia*'s leading edge, however, had locally undermined the boundary layer, and was now letting in a plume of superheated air that was cutting through insulation and working its way toward the inner recesses of the left wing. It is estimated that the plume may have been as hot as 8,000° near the RCC breach. The aluminum support structures inside the wing had a melting point of 1,200°, and they began to burn and give way.

The details of the left wing's failure are complex and technical, but the essentials are not difficult to understand. The wing was attacked by a snaking plume of hot gas, and eaten up from the inside. The consumption began when the shuttle was over the Pacific, and it grew worse over the United States. It included wire bundles leading from the sensors, which caused the data going into the MADS recorder and the telemetry going to Houston to fail in ways that only later made sense. At some point the plume blew right through the top of the left wing, and began to throw molten metal from the insides all over the aft rocket pods. At some point it burned its way into the left main gear well, but it did not explode the tires.

As drag increased on the left wing, the autopilot and combined flight-control systems at first easily compensated for the resulting tendency to roll and yaw to the left. By external appearance, therefore, the shuttle was doing its normal thing, banking first to the right and then to the left for the scheduled energy-management turns, and tracking perfectly down the descent profile for Florida. The speeds were good, the altitudes were good, and all systems were functioning correctly. From within the cockpit the ride appeared to be right.

By the time it got to Texas the *Columbia* had already proved itself a heroic flying machine, having endured for so long at hypersonic speeds with little left of the midsection inside its left wing, and the plume of hot gas still in there, alive, and eating it away. By now, however, the flight-control systems were nearing their limits. The breakup was associated with that. At 7:59:15 Mission Control noticed the sudden loss of tire pressure on the left gear as the damage rapidly progressed. This was followed by Houston's call "And *Columbia*, Houston, we see your tire-pressure messages, and we did not copy your last call," and at 7:59:32 by *Columbia*'s final transmission, "Roger, ah, buh . . ."

The *Columbia* was traveling at 12,738 mph, at 200,000 feet, and the dynamic pressures were building, with the wings "feeling" the air at about 170 mph. Now, suddenly, the bottom surface of the left wing began to cave upward into the interior void of melted and burned-through bracing and structure. As the curvature of the wing changed, the lift increased, causing the *Columbia* to want to roll violently to the right; at the same time, because of an increase in asymmetrical drag, it yawed violently to the left. The control systems went to their limits to maintain order, and all four right yaw jets on the tail fired simultaneously, but to no avail. At 8:00:19 the *Columbia* rolled over the top and went out of control.

The gyrations it followed were complex combinations of roll, yaw, and pitch, and looked something like an oscillating flat spin. They seem to have resulted in the vehicle's flying backwards. At one point the autopilot appears to have been switched off and then switched on again, as if Husband, an experienced test pilot, was trying to sort things out. The breakup lasted more than a minute. Not surprisingly, the left wing separated first. Afterward the tail, the right wing, and the main body came apart in what investigators later called a controlled sequence "right down the track." As had happened with the *Challenger* in 1986, the crew cabin broke off intact. It assumed a stable flying position, apparently nose high, and later disintegrated like a falling star across the East Texas sky.

DIANE ACKERMAN

We Are All a Part of Nature

FROM *PARADE*

A film starts running across my mind's eye, accompanied by the sound of heartbeats and birdsong. It contains my whole experience of Earth, including all the oceans I've floated on or swum under, the skies I've flown through, the lands I've walked upon, the humans and other animals I've known, lots of nature I've never witnessed firsthand but glimpsed in documentaries or read about, and the Earth seen from space.

Naturally, that film would take lifetimes to explore, because nature means the full sum of creation, from the Big Bang to the whole shebang. It includes: spring moving north at about 13 miles a day; afternoon tea and cookies; snow forts; pepper-pot stew; pink sand and confetti-colored cottages; moths with fake eyes on their hind wings; emotions both savage and blessed; tidal waves; pogo-hopping sparrows; blushing octopuses; scientists bloodhounding the truth; memory's wobbling aspic; the harvest moon rising like slow thunder; fat rainbows beneath spongy clouds; tiny tassels of worry on a summer day; the night sky's distant leak of suns; an aging father's voice so husky it could pull a sled; the courtship pantomimes of cardinals whistling in the spring with "what cheer, what cheer, what cheer!"

Sometimes we forget that nature also means us. Termites build mounds; we build cities. All of our being—juices, flesh and spirit—is nature.

Nature surrounds, permeates, effervesces in and includes us. At the end of

our days, it deranges and disassembles us like old toys banished to the basement. There, once living beings, we return to our nonliving elements, but we
still and forever remain a part of nature. Not everyone agrees with me. Many
people harbor an us-against-them mentality in which nature is the enemy and
the kingdom of animals doesn't really include us. Then we can attribute to animals all the things about ourselves that we can't stand.

True, we build more elaborate habitats than other animals who, to the best
of my knowledge, don't require anything like electric cow-milk frothing
machines, beeswax on a flaming string or vaporized flower essence mixed with
musk from the anal sac of civets to encourage breeding. But I could be wrong.
Maybe the wren's liquid melody is equally fantastic. And I'm reluctant to hazard a guess about the necking and petting of alligators, whose cheeks are studded with exquisitely sensitive pleasure nodes. Even at our most domesticated
and tame, we're like pet zebras or grizzly bears—dangerous to anger, always
flirting with a tantrum just under the well-behaved surface. We're remarkable
animals, erudite and loving, but, like circus lions, we will always be wild and
fiercely unpredictable.

Each day, I wake startled to be alive on a planet packed with so much life.
No gasp of sunlight goes unused. Life homesteads every pore and crevice,
including deep dark ocean trenches. Life's rule seems to be variations on every
possible theme: And so we have tree frogs with sticky feet, marsupial frogs, poisonous frogs, toe-tapping frogs, frogs that go peep and many more.

The leafy green abundance we usually think of as nature began with Earth's
earliest life-forms, blue-green algae. Their gift was the cell, a microscopic circus
that still is the basis of a cougar, bombardier beetle and one's nephew. Their
genius was inventing photosynthesis. Around 2.4 billion years ago, they began
building solar power plants under their walls, digesting their surroundings
and, in the process, excreting oxygen, a poisonous gas.

Over time, the algae sheathed the planet, and oxygen fizzed through the
oceans, saturating them. Then the bubbles rose, breathing life into a slaggy sky,
whose cloudbanks thinned as the blues appeared. Hydrogen ballooned away
into space, while heavier oxygen stayed home. Earth became a planet rich in
poisonous, flammable oxygen.

Meanwhile, evolution tinkered with creatures immune to oxygen, including some willing to pool their DNA. Complex animals evolved. And the rest is
history. In every flake of skin, we still resemble those one-celled pioneers. If
they didn't excrete oxygen, we wouldn't be here. So, no matter how politely one
puts it, we owe our existence to the flatulence of blue-green algae. That should

humble us and remind us that we share our origins and future with the rest of life on Earth. We need a healthy environment if we hope to stay healthy.

Most days, I make time to play outside, usually in the garden or on a bike or taking a walk. I live in the country, but nature also means the manicured wilderness of a large city, where flimsy blades of grass crack through cement and fragile snowflakes halt traffic. What feats of strength! A city park lures countless animals from miles away to its bustling green oasis. Surrounded by trees and sky, it's easier to feel a powerful sense of belonging to the pervasive mystery of nature, of being molded by unseen forces older than our daily concerns. Without that, life would feel flat as a postage stamp.

But nature also means comfort, heritage and seasoned home. Indoors, a sensuous activity I heartily recommend is what I think of as "spanieling." Find a shaft of sunlight pouring through a window on a cold day, curl up in the puddle of warmth it creates, relish the breath of sun on your skin and nap with doglike dereliction. If you have trouble turning off your mind-theater, picture yourself as a squirrel, bear or cocker spaniel enjoying a simple sunbath.

Steep yourself in nature. The world will wait.

About the Contributors

DIANE ACKERMAN is the author of twenty works of poetry and nonfiction, including, most recently, *An Alchemy of Mind: The Marvel and Mystery of the Brain; Origami Bridges: Poems of Psychoanalysis and Fire;* and *Animal Sense* (children's), illustrated by Peter Sis. Ms. Ackerman has received many prizes and awards, including a Guggenheim Fellowship, the John Burroughs Nature Award, and the Lavan Poetry Prize. She also has the somewhat unusual distinction of having a molecule named after her—dianeackerone (a crocodilian sex pheromone). She also has hosted a five-hour PBS television series inspired by her book *A Natural History of the Senses.*

She comments: "Although most people feel rivulets of wonder, and even bone-shaking awe, from time to time, not everyone is comfortable expressing those feelings. I suppose what they fear is loss of objectivity. But as so many scientists and artists alike have discovered, life doesn't require you to choose between reason and awe, or between clear-headed analysis and a rapturous sense of wonder. A balanced life includes both. One of the fascinating paradoxes of being human is that we are inescapably physical beings who yearn for transcendence. Unfortunately, language really stumbles when emotions surge. So we don't have a precise vocabulary for complex feelings. Small wonder people resort to metaphors to express their raw joy, mysticism, and sense of wonder."

BARBARA J. BECKER received her PhD in history of science from the Johns Hopkins University. She currently teaches history of science at the University of California, Irvine. For many years, she has actively promoted the use of history in the teaching of both physics and astronomy at the secondary and college levels. With major funding from the National Science Foundation, she developed MindWorks (2000), a series of history-based instructional modules for secondary-school physical science, which is now in use nationally. Her research interests include the role of the amateur in the development of nineteenth-century professional astronomy, the redefining of disciplinary boundaries in the face of new knowledge and new practice, and the role of controversy in shaping the substance and structure of scientific knowledge.

She writes: "Lives are mosaics fashioned out of numerous incremental day to day decisions and happenstance, what historian Pamela Smith has referred to as the 'noise' of a life. Often, in public accounts, these events have been selected and arranged to fit a coherent pattern. It is only by sifting through the archival shards of another's life that we historians accumulate an odd collection of details, and not, I might add, necessarily in the most convenient order. Aided by an abstract form of haptic processing, we manipulate these details in our minds until a vivid three-dimensional image emerges—we see the face, the instrument, the laboratory; we hear the scraping of the observatory dome, the zapping of the Geissler tube; we feel the frustration, the joy, the fear; we smell the battery's noxious fumes, the burning magnesium, the sweet night air; we taste the sweat, the ink, the long-forgotten cup of tea. Astrophysics as we know it today is built on a range of questions and methods that were unimaginable to individuals in the first half of the nineteenth century. It is a hybrid discipline whose emergence and efflorescence required the wholesale restructuring of the boundaries surrounding the theory and practice of astronomy. What made it possible to include such an unorthodox line of investigation within the traditional astronomical community? What prompted others to move into this new thought space? If science is (to its credit) an inertial system, how do changes like this come about? William Huggins's private accounts of his life and work offer us all a glimpse into the story of astrophysics' origins."

MICHAEL BENSON is a widely published writer and award-winning filmmaker. His book *Beyond: Visions of the Interplanetary Probes* won First Prize for Design in the Special Trade General Books Category at the 2004 New York Book Fair and has been called "an aesthetic revelation . . . a spectacular melding of science and art" (*Los Angeles Times*). Benson has contributed articles to the *New York Times* (Op-Ed), *The New Yorker, The Atlantic Monthly,* and *Smithson-*

ian, among other publications. His 1996 film *Predictions of Fire* premiered at the Sundance festival and won several best documentary awards internationally. Benson is currently working on *More Places Forever,* a global road movie.

"As a long-term space exploration enthusiast," he says, "I followed the *Galileo* Jupiter mission almost from the start. The odysseys of our sophisticated, sometimes error-prone, but finally incredibly resilient space robots are mostly fascinating, but *Galileo's* set a new standard. One of the landmark achievements in the history of space exploration, *Galileo* had to overcome numerous potentially mission-terminating problems, but it redeemed itself spectacularly, among other ways by discovering that Jupiter's haunting moon Europa almost certainly has a vast ocean under its fissured ice shell. My *Galileo* article for *The New Yorker* was particularly challenging because the mission's many chapters and achievements had to fit into a comparatively few words."

TOM BISSELL was born in Escanaba, Michigan, in 1974. He attended Michigan State University, served as an English teacher for the U.S. Peace Corps in Uzbekistan, and worked as a book editor for Henry Holt. His fiction and nonfiction have appeared in *Harper's Magazine, McSweeney's, Agni,* the *Boston Review, Granta, GQ, The Believer,* and elsewhere. His first book, *Chasing the Sea,* was published in 2003. He is also the author, with Jeff Alexander, of *Speak, Commentary,* published in 2003, and of a short-story collection, *Death Defier and Other Stories from Central Asia,* to be published in 2005. He is currently writing a book about Vietnam.

"It will probably surprise no one when I say that 'A Comet's Tale' was written by someone fairly deep into a nervous collapse," he says. "Two heady years of nonstop geopolitical trauma had gotten to me, and my thoughts entertained virtually nothing but mortality and meaning and religion and the cosmos. Which is when I got a call from John Jeremiah Sullivan, lately of *Harper's Magazine,* who wanted to know if the topic of the end of the world was of any interest to me. I feel much better now, and I would like to believe that writing 'A Comet's Tale' was of no small aid in that process—even if we are all doomed."

K. C. COLE is the author, most recently, of *Mind Over Matter: Conversations with the Cosmos.* She is a science writer for the *Los Angeles Times* and commentator for Pasadena Public Radio. Other books include *The Hole in the Universe: How Scientists Peered Over the Edge of Emptiness and Found Everything* and *The Universe and the Teacup: The Mathematics of Truth and Beauty.* Cole has written for the *New York Times,* the *Washington Post, Esquire, Newsweek,* and other publications. She has taught at UCLA, Yale, and Wesleyan. Among her recent awards are the American Institute of Physics Science Writing Award and the *Los Angeles Times* awards for deadline reporting and explanatory journalism.

She writes: "This article ends an offer I felt I couldn't refuse: How would I like to come to Fermilab, physicist Janet Conrad asked me, to work on her experiment to search for the sterile neutrino? And so, the following summer, I became the first embedded journalist—so far as I know—on a large-scale physics project. When I wrote about the experience for the *Los Angeles Times*, many readers were surprised to learn that even the most elaborate experiment turns on tiny details and laborious scut work: calibrating phototubes, taping particle counters, finding optimal voltages. Experimental physics, to paraphrase the poet Muriel Rukeyser, is a great deal like real life."

KEAY DAVIDSON was born in Columbia, Georgia, in 1953. He has been a science writer for daily newspapers including the Orlando, Florida, *Sentinel Star* (1979–81), the San Diego County edition of the *Los Angeles Times* (1981–85), the *San Francisco Examiner* (1986–2000), and the *San Francisco Chronicle* (2000–present). He has received the American Association for the Advancement of Science–Westinghouse Award and the National Association of Science Writers' Science in Society Award. He has also written three books, including *Carl Sagan: A Life* (1999), and is finishing a biography of Thomas S. Kuhn.

"I inhabit two contradictory worlds," he comments. "One is the 9:30 to 5:30 world of the newspaper science writer, who celebrates 'breakthrough' research such as the cosmologists' feat depicted in my article. My other world is the post-5:30-and-weekends realm of the book author, an amateur historian-philosopher of science whose ambitions and skepticism exceed those that can be satisfied by deadline writing. As a book writer I try to understand those same scientific circus stunts less reverentially—i.e., to interpret them less as glimpses of a transcendent 'reality' than as projections of ourselves, our society, and our myths . . . as neurological/anthropological constructs, if you will. That galactic superclusters, dark matter, and microwave background radiation really 'exist,' who can doubt? But let's not forget that we perceive them through eyes, occipital lobes, and social epistemologies that are culturally contingent and evolutionarily transient; phlogiston, the Ptolemaic spheres, and the aether once 'existed,' too! Hence the (irritating but inescapable) quotation marks."

ATUL GAWANDE is Assistant Professor of Surgery at Harvard Medical School, Assistant Professor in Health Policy at the Harvard School of Public Health, and a staff writer for *The New Yorker*. He is also the Kessler Scholar at the Center for Surgery and Public Health at Brigham and Women's Hospital in Boston. His book, *Complications: A Surgeon's Notes on an Imperfect Science*, was a finalist for the National Book Award. He and his family live in Newton, Massachusetts.

"Dr. Moore was a looming presence in my training," he explains, "as I alluded to in the essay. He changed medicine fundamentally, but for the hundreds of surgeons who trained in his hospital, he was a moral figure more than anything. He implanted that constant twinge in your conscience that kept you thinking hard about how to help your patients—and turned you around at the hospital exit when you remembered something you hadn't checked. (When an exasperated spouse is waiting outside in the car or you've got a flight you're supposed to catch, that has always been when he mattered most.) So it was a massive shock to learn that he had committed suicide, with a gun no less. At first, I thought this represented a core inconsistency in his life and what he stood for. Then I learned that it didn't."

AARON E. HIRSH is a research fellow in the Department of Biological Sciences at Stanford University. A cofounder of the biotechnology company Inter-Cell, he has since turned his research to basic questions in molecular evolution. Apart from annual pilgrimages to Baja California and Colorado, he lives in Berkeley, where he is currently working on his first book.

He writes: "Perhaps fittingly, the elegant relationship that appeared on Hunter's screen has since given rise to a number of more complex stories—organisms are stubbornly recondite. But Hunter has worked through these complications carefully, and, at least so far, the simple statistical rule seems to persist. The tension between 'reductive' science and the profuse richness of our encounters with nature is palpable in every lecture of the Baja class, and this essay reflects countless siesta-hour conversations with my co-instructors, Graham Burnett, Dmitri Petrov, and Veronica Volny. Subjecting their ideas to Anne Fadiman's brilliant editorial work turned out to be a good strategy for writing an essay. A final thought. 'Signs of Life' touches on two very different ways of engaging the natural world. If we are not careful, one of these will soon be lost to us—even as a possibility. In a world without dragons, one can always collect nudibranchs, but every year in Baja, another species is gone. The Bay of LA, like so many other places, needs protection."

JENNIFER KAHN writes about science and other subjects for *Wired* magazine, where she is a contributing editor, as well as for *Discover, Harper's Magazine,* and *National Geographic.* A graduate of Princeton University and UC Berkeley, she credits John McPhee with saving her from a career as an astrophysicist. From her home in Berkeley, California, she is one block away from a hummingbird nest containing two miniature chicks, which have now been photodocumented more thoroughly than most people's children. She is a recipient of the American Academy of Neurology's 2003 journalism fellowship, and the 2004 CASE-UCLA media fellowship in neuroscience.

Of "Stripped for Parts" she writes, "It was one of those stories that turn out to be dramatically different from the original assignment. Basically, I'd been sent to find out what was new in the world of transplant surgery: the standard 'hooray for scientific progress' tale. Instead, I was struck by how fragile the organ recovery process was—it takes an extremely unusual set of circumstances for someone to die in a way that allows their organs to be salvaged—and how complex. Keeping a donor's body going after brain death is a tricky task, and one that is usually left out of transplant articles. In the end, the piece was quite controversial; there were a lot of angry letters from people who accused me of discouraging donation. The truth is more complex, I think, and my own experience certainly was. Nonetheless, I found the complicated, sometimes unpleasant, details of organ recovery terribly compelling—an aspect of transplantation that not many people seem to know about."

WILLIAM LANGEWIESCHE, a former professional pilot, is a national correspondent for *The Atlantic Monthly*. He is the author of five books: *Cutting for Sign, Sahara Unveiled, Inside the Sky, American Ground,* and *The Outlaw Sea.*

"I had told myself that I would not be doing any more accident reporting," he says, "but the *Columbia* disaster was so clearly interesting and so important to the United States from a policy point of view that I thought I should do it for *The Atlantic Monthly.* What emerged in the investigation was a kind of institutional stupidity of vast proportions. I think many people at NASA—especially among the operational people, the people on the front lines, the astronauts, mission control—felt very strongly the need to face the facts squarely, and that's what Admiral Gehman's CAIB investigation did, with almost brutal frankness."

SUSAN MILIUS, the organismal biology writer at *Science News,* says that she can write about birdsong and butterfly migration and other worthy PG-13 topics for months on end, but her colleagues routinely introduce her as the bug-sex writer. Before this, she lived a publishing-gypsy life, including stints as a part-time copy editor for a scientific journal published in five languages—of which she spoke only one—and a ridiculously miscast phase as a magazine food editor. She was, she admits, a dreadful cook, and one day in the middle of a barbecue seminar in a conference hotel in Kansas City, she realized that what she really should be doing was writing about biology. Even doing so as a science writer for the wire service UPI, the most nerve-fraying job she has had so far, was an improvement over all those half teaspoons and preheated ovens.

"For years," she adds, "I've had to leave out some of the most entertaining stuff when I wrote about this or that research. I had to stay reasonably focused on the findings without indulging too much space in details about the process of an experiment. Yet these details were the ones I repeated to my buddies at lunch: how pollination biologists keep flowers untouched by covering the buds with bridal veiling or how entomologists get their living specimens through airport screening. It was great fun finally to focus on the outtakes."

OLIVER MORTON is a freelance writer and editor who concentrates on scientific knowledge, technological change, and their implications. He is a contributing editor at *Wired,* and his writing has also appeared in *Nature, Science, The New Yorker, National Geographic,* the *New York Times, Discover, Prospect, New Scientist, The American Scholar,* and a range of other newspapers and magazines up to and including *The Hollywood Reporter* (but, alas, only once). His first book, *Mapping Mars: Science, Imagination, and the Birth of a World,* was published in 2002 by Picador (U.S.) and Fourth Estate (UK), and was short-listed for the *Guardian* First Book Award. He is currently working on a book about what photosynthesis means to people, plants, and planets. His Web log can be found at http://mainlymartian.blogs.com.

"Some ideas just make you say wow, and for me the idea of quark nuggets piercing the earth was one of them," he observes. "The idea that there are peculiar particles passing through the earth is a fairly commonplace one in physics, ever since the original searches for solar neutrinos in the 1960s. There are now various experiments around the place that look down *through* the earth for neutrinos coming up from the antipodean sky. But the idea that there might be things which didn't just slip through unnoticed but tore through, making a racket that could be heard by seismometers around the world, struck me as new and fresh and wow-worthy. It also offered a nice counterpoint to asteroid impacts, which are something I've written a fair bit about here and there; here was something similar, in some ways, but which was purely a matter for curiosity, rather than concern. It offered the global scope of a science-fiction disaster movie, but without the disaster.

"Another thing that attracted me to the story was its interdisciplinarity. People endlessly say that science ought to be more interdisciplinary, but in practice making it so can be hard. It's just a lot more convenient to stay in the discipline you started with. In this case, the problem was one that most earth scientists would never have heard of and most physicists didn't find very interesting, because quark nuggets aren't a very hot topic right now. So it was very marginalized—but still the team managed to come together and mount a serious investigation.

"Since the story was written, and partly because of the Indian Ocean earthquake identified by the International Seismological Centre that the story brought to light, the Texas team has decided the October 1993 event is probably terrestrial in origin. But the November event still looks good. They haven't yet had the time or funding to develop a real-time system to look for nuggets as they happen, but it's still something they'd like to do. And Vic Teplitz says one of the detectors currently under construction to watch neutrinos coming up through the earth is interested enough in the idea to be thinking of adding some seismometers to its instrumentation, in case a much smaller quark nugget should wander through.

"In some ways, though, the actual existence or not of the nuggets is a secondary matter, at least to me. The real heart of the story is the inspired idea of using the whole earth as a scientific instrument. If there's a coherent thread to my interests—it's a big if—it's an interest in the many different ways of understanding what planets are and how we relate to them. The idea of a world that is now so caught up in the web of science that we could hear it ringing when flicked by weird cosmic rain was a new way of seeing this planet, and one that spoke to me."

SHERWIN B. NULAND is Clinical Professor of Surgery at the Yale School of Medicine and Fellow of the university's Institution for Social and Policy Studies. He is the author of several well-known books, including *Doctors: The Biography of Medicine* (Alfred A. Knopf, 1988), the story of medicine told in the form of biographies of fourteen of its most prominent contributors; *How We Die* (Alfred A. Knopf, 1994), a *New York Times* best seller that won the National Book Award; *The Wisdom of the Body* (Alfred A. Knopf, 1997, retitled *How We Live* in paperback); *The Mysteries Within* (Simon & Schuster, 2000); *Leonardo da Vinci* (Lipper/Viking Penguin Lives, 2000); *Lost in America,* a memoir (Alfred A. Knopf, 2003); and *The Doctors' Plague* (Atlas/Norton Great Discoveries, 2003). His biography of Moses Maimonides will be published in 2005. He has written for several publications and serves on the editorial boards of *The New Republic, Perspectives in Biology and Medicine,* and *The American Scholar,* for which he wrote a column on medicine entitled "The Uncertain Art," from 1998 to 2003.

"As I approach seventy-four," he says, "I get to the gym about three times a week, pumping iron and dashing along on the treadmill (and concomitantly lusting after the bevy of tightly clad young women who are always there), all of which began six years ago at the urging of the hero of this piece, Manny Papper. Manny had the wisdom to know that the essence of life is how it is lived and not how long. And he knew that it was not only hard work but also fun to strive for the continuing vibrant health that enables us older men and women to enjoy what really counts, which is the love and work that Dr. Freud told us about. To this best friend I've ever had, there was more than theory in the ancient Latin proverb *'dum vivimus, vivamus,'* 'While we live, let us live.'"

DENNIS OVERBYE is a science correspondent for the *New York Times.* Born in Seattle, he majored in physics at MIT before deciding that writing was the only thing he was fit for. His first job in journalism was as a part-time assistant typesetter at *Sky and Telescope* magazine. His articles have appeared in a variety of publications and he is a two-time winner of the American Institute of Physics Science Writing Award. He is the author of *Lonely Hearts of the Cosmos: The Scientific Search for the Secret of the Universe* (HarperCollins, 1991), which was a finalist for the National Book Critics Circle Award, and *Einstein in Love: A Scientific Romance* (Viking, 2000). He lives in Manhattan with his wife, Nancy Wartik, and their daughter, Mira Kamille.

Regarding "One Cosmic Question, Too Many Answers," he writes: "This article is about two of the most vexing questions in science: the nature of the dark energy that is apparently accelerating the expansion of the universe, and the issue of whether physics ever arrives at a unique answer to the question of what is the universe—or as Einstein put it, 'Did God have any choice?' when he created the universe. These two questions came clashing together at a small conference at Stanford in the spring of 2003. I had planned to attend, but wound up in Kazakhstan adopting my daughter instead. But even before I got back I started getting e-mails from various astronomers and physicists, saying that something extraordinary happened.

"In a nutshell, recent developments in string theory—the abstruse, almost mathematically impenetrable, supposed 'theory of everything'—managed to crack a fundamental problem about dark energy that many theorists had regarded as insoluble. But it left them, the theorists, in the uncomfortable position of having to posit literally billions upon billions of different possible universes. If they are right, then most of what we take as the laws of nature are only local effects, like intercosmic weather, subject to the vagaries of some superphysical chance. This has outraged other physicists, who say that Einstein's dream of a final answer is what drives the whole enterprise to begin with. You can't go to a physics conference nowadays without being swept up in dinner-table or coffee-break discussions about this issue. It's a great intersection of science and philosophy."

IAN PARKER is a British journalist who lives in New York. He has been a staff writer at *The New Yorker* since 2000.

"Before I went to Peru," he explains, "Niels Birbaumer had introduced me to people who were almost fully locked-in, but Elías Musiris was the first person I met who had no ability at all to communicate with the outside world. It was very strange to watch the scientist and his subject together: to see Musiris's mind come into view, fleetingly, and see Birbaumer's mood rise and fall—being voluble and indiscreet, Birbaumer is the opposite of locked-in. Soon after he went back to Germany, Birbaumer learned that Musiris's family had lost faith in the TTD method, and the two men are no longer in contact."

KEVIN PATTERSON is an internist practicing on Vancouver Island and in the Canadian Arctic. His fiction and nonfiction have appeared widely in, among others, *The New York Times Magazine*, *Mother Jones*, *Saturday Night*, and the *Globe and Mail*. His short-story collection, *Country of Cold*, was recently published by Nan A. Talese/Doubleday.

Of "The Patient Predator" he writes: "This piece took shape under the encouragement of Clara Jeffrey at *Mother Jones* magazine over the course of about a year. I had been working in the Canadian Arctic for several years and had been thinking about the persistence of tuberculosis there as a reflection of the poverty and isolation of the place. Clara had been interested for a long time in the re-emergence of the disease in New York and throughout inner-city America as a manifestation of globalization and the widening gap between the rich and the poor in that setting. Between these two perspectives 'The Patient Predator' took shape, and works as a sort of synthesis of these points of view. The essential idea emerged that suffering does not localize; that sickness and trouble on the social and geographic margins both moves to the center, and illuminates the center. My thanks to Clara for all her help with this piece. From the writer's point of view, she is a dream."

KAJA PERINA is the editor in chief of *Psychology Today*, for which she hopes to write more frequently. She earned a master's degree from the Columbia University Graduate School of Journalism in 1999. Prior to joining *PT* she was a staff writer for *Brill's Content*.

"I've never made the pilgrimage to Roswell, New Mexico, nor have I watched *The X-Files*," she says. "I was drawn to this story as a way to explore the formation of false memories. What fascinates me about the abduction experience is the tenacity with which abductees cling to it. The researchers were unable to convince a single abductee that the experience is illusory. So you've got this elegant paradigm that shows how these beliefs probably arise, but it is completely useless in the face of subjective experience. The people who could most benefit from an explanation are the ones who vehemently refuse to buy it.

"In fact, they openly disdain it. I wasn't wholly prepared for the hostility towards empirical science that I encountered among intelligent, thoughtful individuals. Because the abduction experience is so integral to their sense of self, they are personally threatened by our attempts to explain it."

MICHAEL POLLAN is a contributing writer for *The New York Times Magazine* and the Knight Professor of Science and Environmental Journalism at UC Berkeley's Graduate School of Journalism. He is the author of *The Botany of Desire* (2001), *A Place of My Own* (1997), and *Second Nature* (1991). He is currently working on a book about the food chains in which humans take part.

"After writing a series of fairly discouraging articles about the industrial food system for *The New York Times Magazine,* I was looking for a more hopeful story to tell about food," he explains. "What I like especially about Slow Food is the way the movement squares the circle of environmentalism and pleasure—linking the defense of biodiversity to decisions as simple, and momentous, as what to have for dinner."

ELIZABETH ROYTE is the author of *The Tapir's Morning Bath: Mysteries of the Tropical Rain Forest and the Scientists Who Are Trying to Solve Them,* which was named a *New York Times* Notable Book of the Year in 2001 (you can learn more about the book at www.tapirsmorningbath.com). Royte has written about science and the environment for *The New York Times Magazine, Harper's Magazine, National Geographic, Outside, Smithsonian, The New Yorker,* and other national magazines. A former Alicia Patterson Foundation Fellow, she is currently working on a book about garbage and hyperconsumption. She lives in Brooklyn with her husband and their daughter.

"Since I reported this story," she writes, "the Environmental Protection Agency has decided not to restrict the use of atrazine, despite several recent studies that show it threatens human health and the environment. The European Union announced it would ban atrazine, and the NRDC is suing the EPA for failing to protect endangered species from the herbicide; for violating the Freedom of Information Act by withholding evidence that the pesticide industry may have undue influence over federal health standards for atrazine; and for failing to consider cancer risks appropriately. Tyrone Hayes, who is now a full professor at Berkeley, continues to study the environmental impacts of atrazine."

Tom Siegfried was born in Ohio and migrated to Texas, graduating from Texas Christian University in 1974 with majors in journalism, chemistry, and history. He earned a master's degree from the University of Texas in 1981. He joined the *Dallas Morning News* in 1983 and has been science editor there since 1985. He has written two popular science books: *The Bit and the Pendulum* (2000) and *Strange Matters* (2002). His work has been recognized with awards from the American Chemical Society, the American Psychiatric Association, and the American Association for the Advancement of Science.

He writes: "Outside economics, game theory is a rather obscure field of research, although it's well known to some biologists. My interest in it grew from a realization that John Nash's fifty-year-old math had been showing up in all sorts of fields lately, far from its original applications, especially in neuroscience. There will be much more to be said and written about game theory in the years ahead, especially, I believe, regarding its relationship to statistical physics and the mathematics of social interaction. Game theory, in fact, may turn out to be the glue that unites physics, economics, biology and sociology."

Neil DeGrasse Tyson was born and raised in New York City, where he was educated in the public schools clear through his graduation from the Bronx High School of Science. Tyson went on to earn his BA in physics from Harvard and his PhD in astrophysics from Columbia University. In addition to dozens of professional publications on galaxies, Dr. Tyson has written, and continues to write, for the public. Since January 1995, he has written a monthly column, called "Universe," for *Natural History* magazine. In addition to a memoir, *The Sky Is Not the Limit: Adventures of an Urban Astrophysicist,* Tyson has written six books on the universe, the latest of which is *Origins: Fourteen Billion Years of Cosmic Evolution,* coauthored with Donald Goldsmith.

"I wrote 'Gravity in Reverse' as one in a series of essays for *Natural History* magazine in which I profiled the birth and evolution of the universe," he explains. "For this particular subject, I felt compelled to speak candidly about our cosmic ignorance. Among all the unknowns in modern astrophysics, the problems of dark matter and dark energy loom heaviest on our collective psyche. We simply do not yet know what 93 percent of the universe is made of. Living in this darkness, one cannot help wonder whether modern astrophysics sits at the doorstep of a major breakthrough that will transform our abject ignorance into profound insight. Such was the case a century ago, when the resolution of some long-standing problems in classical physics led to the discovery of quantum mechanics and relativity, two remarkable new ways to look at the world."

JOHN UPDIKE was born in Shillington, Pennsylvania, in 1932. He attended the local pubic schools, then Harvard College and the Ruskin School of Drawing and Fine Art, in Oxford, England. From 1955 to 1957 he was a member of the staff of *The New Yorker,* to which he is a frequent contributor. Since 1957 he has lived in Massachusetts, as a freelance writer. His novels have won the Pulitzer Prize, the National Book Award, the National Book Critics Circle Award, and the Howells Medal.

"As a poet, and especially as a light-verse writer," he writes, "I have kept my eye on the heavens and the quirky discoveries of science. The bright nearness of Mars deserved a verse or two, I thought, and what came out isn't as light as I thought it would be. It made *The New York Times Book Review,* and now this."

JOHN NOBLE WILFORD, a senior science correspondent for the *New York Times,* has for more than forty years reported on space exploration and astronomy, paleontology and archaeology—the long-ago and faraway, as he puts it. He has taught science writing at Princeton and the University of Tennessee and was editor of the newspaper's science staff at the creation of its weekly section Science Times. His honors include a Pulitzer Prize in 1984 and another in 1987 that he shared for coverage of the aftermath of the space shuttle *Challenger* disaster. He is the author or editor of nine books, including *The Mapmakers* and *The Riddle of the Dinosaur.*

"When the archaeologist Katharina Galor alerted me to plans for a symposium on Wadi Arabah," he recollects, "my first reaction was not unusual, or unreasonable. I asked, where is Wadi Arabah, and why is anyone interested? Many are the times that I start my pursuit of an archaeological story with a remedial lesson in geography and history. It didn't take long this time to learn that the story had some irresistible elements, not the least an association with Lawrence of Arabia, who once crossed the forbidding wadi that many thought to be virtually uncrossable. The story of this physical divide between Israel and Jordan also offered a glimmer of hope in the otherwise bleak geopolitical landscape of the Middle East. The wadi was a place and a subject of research on which Israelis and Jordanians appeared to be ready to join forces, and this was newsworthy and to be encouraged. So I went to the symposium. I was reminded once again that archaeology is not all ancient history. Archaeologists may study the past, as I wrote, but they live in the present."

Acknowledgments

A Note from the Series Editor

Submissions for next year's volume can be sent to:

Jesse Cohen
c/o Editor
The Best American Science Writing 2005
Ecco/HarperCollins
10 East 53rd Street
New York, NY 10022

Please include a brief cover letter; manuscripts will not be returned. Submissions made electronically are also welcomed and can be e-mailed to jesseicohen@netscape.net.

Thought-Provoking and Exciting Scientific Inquiry

An annual series dedicated to collecting the best science writing of the year from the most prominent thinkers and focusing on the most current topics in science today.

THE BEST AMERICAN SCIENCE WRITING 2004
Edited by Dava Sobel • ISBN 0-06-072640-7 (paperback) • ISBN 0-06-072639-3 (hardcover)
The fifth volume in *The Best American Science Writing* series is edited by the bestselling author of *Longitude* and *Galileo's Daughter*, Dava Sobel. Essays by Oliver Sacks, Sherwin Nuland, Michael Pollan, John Updike, Atul Gawande, and many others cover the full range of scientific inquiry—from biochemistry, physics, and astronomy to genetics, evolutionary theory, and cognition.

THE BEST AMERICAN SCIENCE WRITING 2003
Edited by Oliver Sacks • ISBN 0-06-093651-7 (paperback) • ISBN 0-06-621163-8 (hardcover)
Oliver Sacks, one of the foremost thinkers/writers on neurology and medicine, and the bestselling author of *The Man Who Mistook His Wife for a Hat* and *Awakenings,* edits this fourth installment of the annual series. Featuring articles from: Peter Canby, Charles C. Mann, Atul Gawande, Liza Mundy, Floyd Skloot, Frank Wilczek, Marcelo Gleiser, Natalie Angier, Margaret Wertheim, Jennifer Kahn, Michelle Nijhuis, Gunjan Sinha, Trevor Corson, Siddhartha Mukherjee, Michael Klesius, Susan Milius, Thomas Eisner, Lawrence Osborne, Brendan I. Koerner, Joseph D'Agnese, Danielle Ofri, Roald Hoffmann, Leonard Cassuto, Dennis Overbye, Richard C. Lewontin, and Richard Levins.

THE BEST AMERICAN SCIENCE WRITING 2002
Edited by Matt Ridley • ISBN 0-06-093650-9 (paperback)
Edited by the renowned and bestselling author of *Genome,* Matt Ridley, the third volume in the series includes pieces by: Lauren Slater, Atul Gawande, Lisa Belkin, Margaret Talbot, Sally Satel, Jerome Groopman, Gary Taubes, Joseph D'Agnese, Christopher Dickey, Michael Specter, Mary Rogan, Sarah Blaffer, Natalie Angier, Julian Dibbell, Carolyn Meinel, David Berlinksi, Tim Folger, Oliver Morton, Steven Weinberg, Nicholas Wade, and Darcy Frey.

THE BEST AMERICAN SCIENCE WRITING 2001
Edited by Timothy Ferris • ISBN 0-06-093648-7 (paperback)
Pulitzer Prize and National Book Award nominee Timothy Ferris, one of the preeminent writers about astronomy, edits this second volume of outstanding science writing. The contributors include: John Updike, Michael S. Turner, Natalie Angier, Joel Achenbach, Erik Asphaug, John Archibald Wheeler, Stephen S. Hall, Richard Preston, Peter Boyer, John Terborgh, James Schwartz, Ernst Mayr, Greg Critser, Andrew Sullivan, Malcolm Gladwell, Helen Epstein, Debbie Bookchin and Jim Schumacher, Stephen Jay Gould, Tracy Kidder, Jacques Leslie, Robert L. Park, Alan Lightman, and Freeman Dyson.

THE BEST AMERICAN SCIENCE WRITING 2000
Edited by James Gleick • ISBN 0-06-095736-0 (paperback)
This first volume in an annual series carries the imprimatur of Pulitzer Prize nominee James Gleick, the celebrated chronicler of scientific social history. This stellar collection includes the writings of James Gleick, George Johnson, Jonathan Weiner, Sheryl Gay Stolberg, Deborah M. Gordon, Francis Halzen, Timothy Ferris, Stephen S. Hall, Floyd Skloot, Denis G. Pelli, Douglas R. Hofstadter, *The Onion,* Don Asher, Natalie Angier, Stephen Jay Gould, Susan McCarthy, Peter Galison, and Steven Weinberg.